Lecture Notes in Mathematics 1597

Editors:
A. Dold, Heidelberg
B. Eckmann, Zürich
F. Takens, Groningen

Bernd Herzog

Kodaira-Spencer Maps in Local Algebra

Springer-Verlag

Berlin Heidelberg New York
London Paris Tokyo
Hong Kong Barcelona
Budapest

Author

Bernd Herzog
Matematiska Institutionen
Stockholms Universitet
S-113 85 Stockholm, Sweden
E-mail: herzog@matematik.su.se

Mathematics Subject Classification (1991): 13D40, 13D10, 14B12, 16S80

ISBN 3-540-58790-X Springer-Verlag Berlin Heidelberg New York

CIP-Data applied for

© Springer-Verlag Berlin Heidelberg 1994
Printed in Germany

Typesetting: Camera-ready T_EX output by the author
SPIN: 10130247   46/3140-543210 - Printed on acid-free paper

Kodaira-Spencer maps in local algebra

Contents

Introduction

One of the most famous results in commutative algebra is Serre's theorem asserting that the localizations of a regular local ring are again regular. A quantitative refinement of this assertion, proved by Nagata [Na55] in 1955, states that for every local ring A and every prime ideal P of A such that A/P is analytically unramified, there is the following inequality between the multiplicities of the ring A and its localization A_P.

$$(1) \qquad\qquad e_0(A_P) \leq e_0(A)$$

Already Nagata wondered whether the assumption about A/P to be analytically unramified is necessary and hinted at the fact that one could skip it, if one were able to prove an inequality for the multiplicities of certain special flat couples of local rings. In 1959, C. Lech [Le59] suggested that one should try to prove this inequality for arbitrary flat couples, i.e.

$$(2) \qquad\qquad e_0(A) \leq e_0(B)$$

for arbitrary flat local homomorphisms $f : A \to B$ of local rings (A, m) and (B, n). To support his suggestion, C. Lech proved this inequality for first special examples. As a further refinement he asked in 1964 whether it is possible to prove even the corresponding inequality between sum transforms of the associated Hilbert functions, i.e., to prove the existence of a non-negative integer i such that

$$(3) \qquad\qquad H_A^i(n) \leq H_B^i(n)$$

for every n. Lech found out that the inequality holds always for $n = 1$ (see also [Va67]) and announced it for arbitrary n, if the special fibre B/mB of $f : A \to B$ is a complete intersection ([Le?], see also [He90]). In 1970, Hironaka [Hi70] asked whether it is possible to prove the inequality even for $i = 1$. He treated the case that the special fibre is a hypersurface and used his result to show that the local Hilbert function cannot increase under permissible blowing up.

The above inequalities (2) and (3) are in fact assertions of local deformation theory. In a more geometric setting they mean that given a flat morphism $f : (X, x) \to (Y, y)$ of locally Noetherian schemes (X, x), (Y, y), i.e., a *deformation germ*, there is the inequality

$$(4) \qquad\qquad e_0(Y, y) \leq e_0(X, x)$$

for the multiplicities at $y \in Y$ and $x \in X$ and, respectively, the inequality

$$(5) \qquad\qquad H_{(Y,y)}^i \leq H_{(X,x)}^i$$

for the associated Hilbert series. The latter inequality means that all pairs of corresponding coefficients should satisfy the inequality. From the point of view of deformation theory it is quite natural to start with a fixed *local singularity* (i.e., a local scheme (X_0, x_0)) and to ask whether the *deformations* of this singularity (X_0, x_0), i.e., the flat morphisms $f : (X, x) \to (Y, y)$ with *special fibre* $(X_y, x) := f^{-1}(y)$ equal to (X_0, x_0), satisfy Lech's inequality. In this terminology Lech's main result ([Le?],[He90]) states that every deformation of a complete intersection satisfies (5) with $i = 1$. The weaker inequality (4) was established by Lech already in 1959 for the deformations of thick points with local ring

$$K[X_1, \ldots, X_N]/(X_1, \ldots, X_N)^d$$

(where $N \geq 2$) and for the deformations of arbitrary complete intersections. In his paper [He91] the author could show that inequality (5) holds even with $i = 0$ for many other special fibres, but in general he could derive only the inequality

(6) $$H^1_{Y,y} \cdot H^0_{X_y,x} \geq H^1_{X,x}$$

(without the flatness assumption, see [He82]) which is, in some sense, converse to Lech's one. It is natural to ask for conditions ensuring that equality holds in (6) (in which case (5) is trivially satisfied). One such even equivalent condition is the flatness of the morphism $df : C(X, x) \to C(Y, y)$ induced by f on the tangent cones at x and y, respectively. The morphism f is called *tangentially flat* in this case (*graduationally flat* in [He82]). Moreover it is possible to give sufficient conditions for singularities having the property that all their deformations are tangentially flat. One such condition is that Schlessiger's module T^1 of the tangent cone shouldn't have (non-zero) homogeneous elements of degree less than -1 (see [He91]). The condition is fairly easy to verify in many situations, is even necessary for homogeneous singularities, and enables one to find many classes of singularities such that all deformations are tangentially flat and therefore satisfy (5) with $i = 0$ (see [He91]).

Another approach to formula (1) is to find a direct refinement in terms of local Hilbert functions. Such a refinement is the following inequality.

(7) $$H^{i+1}_{A_P}(n) \leq H^i_A(n)$$

(n an arbitrary non-negative integer). This inequality was proved by Lech [Le64] in the case of excellent local rings with $i = 1$ (see also [Be70]). If one uses results of B. Singh [Si74] on the behavior of local Hilbert functions under permissible blowing up, Bennett's proof gives inequality (7) with $i = 0$. Note that the gap in Singh's paper (for the proof of the main theorem, Lemma (4.5) is insufficient) can be filled (see [He80'], Lemma (2.4)).

In a remarkable joint paper with T. Larfeldt [L-L], C. Lech proved in 1981 that the two inequalities (3) and (7) are equivalent. However, the fact that (7) is known in the geometric (i.e., excellent) case has no implication with respect to (3). It is not difficult to see that inequality (3) is much closer to geometry than

(7). In particular, (3) is easily reduced to the case of Artinian local rings. So the result of Larfeldt and Lech states, roughly speaking, if something is true in the non-geometric (i.e., non-excellent case), then this has consequences for geometry. This means, to the author's opinion, that either Lech's inequality doesn't hold in the general case or the inequality reflects a rather deep property of singularities related with an anyhow mysterious connection between the non-geometric part of commutative algebra and infinitesimal properties of algebraic varieties.

In case the inequality doesn't hold in general, it will be hard to find a counterexample, for, in view of the above mentioned cases, when the inequality is known, one has to find a fairly small locus on the Hilbert scheme. The formal versal deformations of the most simple singularities, which are not known to satisfy Lech's inequality depend on about 60 variables. The equations defining the base ring of the formal versal deformation (considered up to degree 3, - the degree 2 situation is the first interesting one) fill many pages so that it is hopeless to find a reasonable specialization. Therefore, the best what one can currently do is to look for larger and larger classes of singularities satisfying Lech's inequality and hope that one can get this way some idea of the singularities which might fail to do so. The purpose of the material presented here is to contribute to this aim. More precisely, we want to prove the following weakened version of Lech's inequality.

$$(8) \qquad \mathrm{H}^1_{Y,y} \cdot \mathrm{H}^0_{X_y,x} \leq \mathrm{H}^1_{X,x} \cdot \prod_{d=2}^{\infty} \left(\frac{1 - T^d}{1 - T} \right)^{n(d)} .$$

for every residually separable flat morphism $f : (X, x) \to (Y, y)$ of germs (X, x), (Y, y) of locally Noetherian schemes with special fibre $(X_y, x) := f^{-1}(y)$. Here

$$n(d) := \dim \mathrm{T}^1_{C(X_y, x)}(-d)$$

is the dimension of the degree $-d$ part of Schlessinger's T^1 associated with the tangent cone $C(X_y, x)$ of the special fibre.

Formula (8) may be interpreted as a numerical refinement of the criterion [He91], Th. 2.5, characterizing the singularities having exclusively tangentially flat deformations. Contrary to (6), the estimation (8) is in the same direction as Lech's inequality and allows often to prove inequalities of type (4) and (5), particularly for singularities with few elements of negative degree in Schlessinger's T^1.

To prove (8), we have to study the morphism $\phi : (Y, y) \to (M, *)$ that defines f as the result of a base change from the formal versal family of (X_y, x), and the mapping

$$d\phi : C(Y, y) \to C(M, *)$$

induced by ϕ on the tangent cones at the base points. The latter morphism may be considered as a local analogue of the Kodaira-Spencer mapping which plays an important role in the deformation theory of compact complex manifolds (see [Ko86]). In proving (8), the morphism $f : (X, x) \to (Y, y)$ is modified via base change of a very special type that allows one to control the behavior of the Hilbert series and that decreases the negative part of the image of $d\phi$. In case the image of this negative part is sufficiently small, f is proved to be tangentially flat, i.e., (8) is trivially true, and the general case is treated inductively depending upon the size of this image. Thus the proof heavily depends upon a good description of $\mathrm{Im}(d\phi)$, and this is the place where the theorems on tangential flatness are needed in a generalized version. The natural filtration of the local ring (A, m) given by the powers of the maximal ideal must be replaced by a more general type of filtration $F_A := (F_A^d)_{d \in \mathbb{N}}$ where the ideals F_A^d are such that A/F_A^d is Artinian and such that $F_A^d \cdot F_A^{d'} \subseteq F_A^{d+d'}$.

More accurately, one should study the Kodaira-Spencer mapping associated not with the morphism $f : (X, x) \to (Y, y)$ but the one coming from the induced map $df : C(X, x) \to C(Y, y)$ on the tangent cones. This corresponds to the phenomenon that it is Schlessinger's T^1 of the tangent cone $C(X_y, x)$ and not of the fibre X_y itself, which decides whether X_y has tangentially non-flat deformations. But the term "Kodaira-Spencer mapping of df" doesn't make sense at first glance, since df is almost never flat, so there is no base change morphism giving df as a result of base change from a universal family. However, there are many possibilities to construct induced morphisms analogous to df which correspond to certain pairs of filtrations on the local rings at x and y, and many of these filtrations define flat morphisms of cones, hence are suited to construct a Kodaira-Spencer map. So the problem is to find a canonical pair of such filtrations. It turns out that among the filtrations giving flat morphisms of cones there is a minimal pair, and this is our candidate. Almost nothing is known about these minimal filtrations currently, and possibly they are rather exotic in certain cases. So we have to reprove the usual theorems on tangential flatness (see [He82] and [He91]) for the case of filtered local rings.

Most of these theorems have analogues in the yet more general context of filtered modules as one can see from an early version of this monograph [He92]. Since the module situation is extremely technical, we spent much care to get rid of it. Up to a few relicts, the present treatment states the results for filtered local rings where the filtrations satisfy an Artin-Rees type condition. The formulations of the theorems and their proofs are much simpler this way. We hope this will make the ideas behind them more transparent to the reader. In the classical (non-filtered) case, the proofs used, as a basic tool, a relative variant [He91, Proposition (1.6)] of Cohen's theorem that each complete local ring is isomorphic to a factor of a formal power series ring. Unfortunately we couldn't prove a general filtered version of it. This is finally the reason that we originally formulated the theorems on filtered tangential flatness in the language of modules where Cohen factorization translates into the trivial fact that each filtered module is the homomorphic image of a "tangentially free" one. Since Cohen factorization is quite useful when dealing with examples, we include a variant of the theorem for quasi-homogeneous filtrations with positive real weights (see 5.9).

In Chapters 1 to 6 we reprove the elementary part of [He91] (covering [He82] and [He83]) in the new context. Using these results, it is possible to prove also a generalized version of the main theorem in [He91] characterizing singularities with tangentially flat deformations in terms of Schlessinger's T^1 (or, equivalently, in terms of the normal module of the fibre with respect to some formal embedding into a regular scheme). The reader is referred to [Ja90], where this generalized criterion is proved and used to construct new fibres such that Lech's inequality holds.

Chapter 1 contains the most important definitions and a few elementary results on the filtrations we are dealing with. To motivate the later restriction to Artin-Rees filtrations of local rings, we give a characterization of these filtrations in terms of the associated complete rings (see 1.17).

In Chapter 2 we prove variants of well-known lemmas which will be frequently used later. The material is present mainly for reference purposes. The reader might skip it and return to it when appropriate.

Chapter 3 is devoted to the fact that tangential flatness is preserved under surjective base change (see 3.8) and its corollaries. The proofs are along the lines of M. Brundu's paper [Br85], where the case of I-adic filtrations is treated. Further we introduce the basic exact sequence (see 3.14) which is used later to establish the connection of tangential flatness to Hilbert series, and we spend some time to prove a kind of inverse (see 3.15) to the base change theorem 3.8. The analogue 4.4 of Brundu's Main theorem [Br85, see (3.3) and (3.4)] is proved in the next chapter using arguments essentially different from Brundu's. The reason is that we don't understand Brundu's argument in the proof of her Theorem (3.3) claiming that her isomorphism $\beta : S \otimes_A B \to T$ is in fact an isomorphism of filtered rings. So we decided to chose a different (more difficult) approach.

In Chapter 5 we introduce the notion of distinguished basis, which is, in some sense, an analogue in the situation of flat local extensions of the notion of a free generating set and which gives us the possibility to introduce "structure constants". The main result 5.8 in this chapter characterizes tangential flatness in terms of these structure constants. When applied to the versal deformation of a (local) singularity it says that the set of quasi-homogeneous filtrations (with positive real weights) having the property that all deformations in an appropriate filtered category are tangentially flat is a finitely generated convex polyhedron. Unfortunately this polyhedron seems to depend heavily upon the local embedding of the singularity into a non-singular scheme. Nevertheless we expect that it will play an important role in the context of Lech's conjecture.

Chapter 6 relates tangential flatness with properties of Hilbert series. The usual theorems known from [He82] are generalized to the filtered case. The proofs differ somewhat from and are more complicated than the proofs known for the natural filtrations. One has to avoid certain enumerative arguments comparing dimensions of vector spaces, which may be infinite in the general case, and has to work instead with exact sequences. But possibly the deductions are even more natural now; the results are the same. We have added two easy but useful theorems on the composition of tangentially flat morphisms, which we learned about from late Christer Lech in a private communication.

In Chapter 7 we introduce the notion of flatifying filtration for a homomorphism of filtered local rings, which is defined as a filtration on the base ring containing the given filtration and making the given homomorphism tangentially flat. We prove the existence of a unique minimal flatifying filtration. The hard part of the proof is to show that this filtration has the Artin-Rees property (see 7.9). As an interesting side result we get that each naturally filtered flat local homomorphism becomes tangentially flat by a base change of a very special type (see 7.8).

Schlessinger's T^1 and the Kodaira-Spencer mapping are constructed in Chapter 8. The central result here is that the minimality property of a flatifying filtration implies that an associated Kodaira-Spencer mapping is injective in certain negative degrees (see 8.13). Contrary to earlier announcements we cannot establish this result for general local homomorphisms, but must restrict to residually rational ones. This is finally the reason that our main result 9.2 is valid for residually separable homomorphisms only. We spend some care to clarify the reason why we can't prove the general statement, showing that the Kodaira-Spencer map is always injective in appropriate degrees when restricted to a certain subspace (see 8.13). One interpretation of our difficulties in the general case is that the minimal flatifying filtration might contain too much information about the extension of the residue class fields induced by the given morphism f whereas the Kodaira-Spencer map itself forgets all such information (see 8.9(iv)).

In Chapter 9 we use the results obtained so far to prove inequality (8). In a first step we establish an inequality of the indicated type with Schlessinger's T^1 replaced by data depending upon the minimal flatifying filtration of the given homomorphism (see 9.1). The final proof of (8) is essentially a comparison of these data with Schlessinger's T^1 based on the injectivity result of Chapter 8. Since the latter is available only for residually rational homomorphisms, we must first reduce the proof of (8) to this special case, which is done using a standard argument of Cohen structure theory. We conclude Chapter 9 with examples illustrating our result and a list of problems related with Lech's inequality and tangential flatness.

The last chapter has the character of an extended example. We study the germ of the universal family at a fixed point of the Hilbert scheme and describe the minimal flatifying filtration on the local ring at the given point. We use this description to give a simplified proof of the main result from [He91] (in the residually rational case) and to show that the question whether a tangentially flat homomorphism can be lifted preserving tangential flatness depends upon finitely many obstructions. Implicitly the considerations of this chapter show that there is a relation between the minimal flatifying filtration on the local rings of the Hilbert scheme and the graded structure of these rings introduced by Pinkham [Pi74] in the homogeneous case. Since this has implications for Lech's inequality, we will come back to this point in a later publication.

Acknowledgments

I want to express my gratitude to Jan-Erik Roos (Stockholm), Ralf Fröberg (Stockholm), and Jörg Jahnel (Göttingen) for their interest in the topic and for many stimulating discussions, which have activated my efforts and opened a wide perspective to me for further work in the area. I also want to thank Jan-Erik Roos for having introduced me to the Bayer-Stillman computer algebra system Macaulay, and both, Jan-Erik Roos and Torsten Ekedahl (Stockholm), for writing a Macaulay script that calculates the generator degrees of the graded normal module for all singularities in embedding dimension 4 such that the tangent cone is defined in degree 2. Moreover, I am obliged to Torsten Ekedahl for his patient help when I had trouble with the Sparc stations of the institute. I am particularly obliged to Gert Almkvist (Lund) who encouraged me to work on the material presented here and supplied me with mathematical literature over many years. Last not least I want to thank Gabriele Dietrich (Jena) who went carefully through the manuscript, discovered many misprints and errors, and suggested a substantial simplification in the proof of 6.12. When I wrote the manuscript, I was partially supported by the Swedish Natural Science Research Council.

Notation

Throughout we use the conventions of commutative algebra as in [Ma86] or [Bo61]. Local rings will be always considered to be Noetherian.. A local homomorphism of local rings is called *residually rational*, if the induced homomorphism of the residue classes is an isomorphism. Similarly, it is called *residually separable*, if it induces a separable field extension. Given two submodules N_1 and N_2 of the module M, we will write

$$(N_1 \bmod N_2)$$

to denote the canonical image $(N_1 + N_2)/N_2 \subseteq M/N_2$ of N_1 in M/N_2. The terminology below will be used frequently, often without any further explanation.

$\mathbb{N}$	non-negative integers
$\mathbb{Z}$	integers
$\mathbb{Q}$	rational numbers
$\mathbb{R}$	real numbers
$\mathbb{C}$	complex numbers
$< a, b >$	the scalar product $\sum_{\lambda \in \Lambda} a_\lambda b_\lambda$ of the families $a := (a_\lambda)_{\lambda \in \Lambda}$ and $b := (b_\lambda)_{\lambda \in \Lambda}$ indexed by one and the same set, where one of the families should have only finitely many non-zero elements and the elements of both families are from appropriate rings or modules such that multiplication and summation is possible.
$\Omega(A)$	set of maximal ideals of the ring A
$m(A)$	maximal ideal of the local ring A
$L_A(M)$	length of the module M over the ring A
$M^{(\Lambda)}$	direct sum of copies of the module M where the number of copies is equal to the cardinality of the set Λ
$\mu_A(M)$	minimum number of generators of the module M over the ring A
F_M^d	d-th filtration submodule of the filtered module M
$M(d)$	subgroup of homogeneous elements of degree d, if M is a graded module, or M/F_M^{d+1} in case M is a filtered module
$M(\le e)$	direct sum of all homogeneous parts $M(d)$ with $d \le e$ of the graded module M

$M(\geq e)$	direct sum of all homogeneous parts $M(d)$ with $d \geq e$ of the graded module M
$\mathrm{G}(M)$	graded module associated with the filtered module M, see 1.13
$\mathrm{G}_F(M)$	same as $\mathrm{G}(M)$ where the A-module M is considered to be filtered with respect to the filtration F (in case F is a filtration of M) and FM (in case F is a filtration of A), respectively. In the second case FM denotes the filtration of M generated by F, see 1.13
$\cap F_M$	intersection of all filtration submodules of the filtered module M
$\mathrm{ord}_M(x)$	$= \sup\{d \in \mathbb{N} \mid x \in F_M^d\}$, the order of the element x in the filtered module M. Sometimes, M will be a factor module of the module containing x. In this case the symbol denotes the order of the residue class in M of the element x.
$M^\wedge$	completion of the filtered module M, see 1.16
$\mathrm{cl}(N)$	closure of the submodule $N \subseteq M$ in the completion of the filtered module M, see 1.16
$\mathrm{R}_B(x)$	module of syzygies of the family x over the ring B, see 3.3
H_M	Hilbert series of the graded module M or the filtered ring $M = A$, see 6.1 and 6.3, respectively.
H_M^i	i-th sum transform of the Hilbert series of the graded module M or the filtered ring $M = A$, see 6.1 and 6.3, respectively.
$\mathbb{P}_K^N$	projective N-space over the field K
$\mathbb{H}$	Hilbert scheme of length n subschemes in $\mathbb{P}_K^N$ where n is a fixed positive integer
$\mathbb{U} \to \mathbb{H}$	the universal family of the Hilbert scheme $\mathbb{H}$

1. Ring filtrations

Remark

This chapter contains the most important definitions and a few elementary results on the filtrations we are going to deal with. Artin-Rees filtrations are defined in 1.12 and, in the case of local rings, characterized in terms of the associated complete rings in 1.17(iii).

1.1. Graded rings

A *graded ring* is defined to be a ring G which admits a direct sum decomposition

$$G = \bigoplus_{d=0}^{\infty} G(d)$$

into submodules $G(d)$ over the ring $G(0)$ such that

$$G(d) \cdot G(d') \subseteq G(d + d')$$

for arbitrary $d, d' \in \mathbb{N}$. The elements of $G(d)$ will be called *homogeneous* of *degree* d, and we will write

$$\deg(x) = d$$

to indicate that $x \in G$ is homogeneous of degree d, i.e., $x \in G(d)$. Note that the zero element may have any degree. The ideal

$$G^+ := \bigoplus_{d=1}^{\infty} G(d)$$

is called the *irrelevant ideal* of G. A *graded module* over the graded ring G is a module M over G that decomposes into a direct sum,

$$M = \bigoplus_{d=d_0}^{\infty} M(d)$$

$(d_0 \in \mathbb{Z})$ of submodules $M(d)$ over $G(0)$ satisfying

$$G(d) \cdot M(d') \subseteq M(d + d')$$

for arbitrary $d \in \mathbb{N}, d' \in \mathbb{Z}$. We will use the convention that

$$G(d) := 0 \text{ for } d < 0 \text{ and } M(d) := 0 \text{ for } d < d_0.$$

Given a graded module M over the graded ring G and an integer e, the module

$$M[e]$$

is defined to be the graded module over G such that

$$M[e](d) := M(d + e)$$

for every $d \in \mathbb{Z}$, the module operations being the same as for M. It is called the module obtained from M by *shifting degrees* e times to the left (or $-e$ times to the right). A *graded ring homomorphism* is a ring homomorphism $f : G \to G'$ with G and G' graded rings such that $f(G(d)) \subseteq G'(d)$ for every $d \in \mathbb{N}$. A *graded module homomorphism* over the graded ring G is a G-linear map $f : M \to M'$ of graded G-modules M and M' such that $f(M(d)) \subseteq M'(d)$ for every $d \in \mathbb{Z}$.

1.2. Filtered rings

Let A be a ring. A *ring filtration* of A is a family

$$F := (F^d)_{d \in \mathbb{N}}$$

of ideals $F^d \subseteq A$ such that the following conditions are satisfied.

(i) $F^d \supseteq F^{d+1}$

(ii) $F^0 = A$

(iii) $F^d \cdot F^{d'} \subseteq F^{d+d'}$

Later we will also define module filtrations. Nevertheless we will sometimes use the term *filtration* for a ring filtration when it is clear from the context which kind of filtration we have in mind. Further we will use the following terminology.

$$F^d = A \text{ for } d < 0, \quad F^{-\infty} = A, \quad F^\infty := \bigcap_{d=0}^{\infty} F^d$$

$$G(A) := G_F(A) := \bigoplus_{d \in \mathbb{N}} G_F(A)(d), \quad G(A)(d) := G_F(A)(d) := F^d / F^{d+1}$$

Note that condition (iii) above implies that $G(A)$ has the structure of a graded ring with multiplication

$$(a + F^{d+1}) \cdot (a' + F^{d'+1}) := (aa' + F^{d+d'+1}) \quad \text{for } a \in F^d,\ a' \in F^{d'}.$$

The ring $G(A)$ is called the *associated graded ring*. A *filtered ring* is a ring A equipped with a ring filtration F_A. Sometimes we will refer to the pair (A, F_A) as to a filtered ring in order to make explicit which filtration of A we have in mind. Given a filtered ring A, the *order* of an element $a \in A$ is defined as

$$\operatorname{ord}(a) = \operatorname{ord}_A(a) = \operatorname{ord}_{F_A}(a) := \sup\{d \in \mathbb{N} | a \in F_A^d\}$$

A *homomorphism of filtered rings* is a ring homomorphism $f : A \to B$ such that A and B are filtered rings and such that $f(F_A^d) \subseteq F_B^d$ for every d. Note that such a homomorphism induces a *homomorphism of the associated graded rings*,

$$G(f) : G(A) \to G(B), \quad a + F_A^{d+1} \mapsto f(a) + F_B^{d+1}.$$

A homomorphism $f : A \to B$ of filtered rings is called *tangentially flat* if $G(f)$ is a flat ring homomorphism, i.e., if the functor

$$\otimes_{G(A)} G(B) : \mathrm{Mod}_{G(A)} \to \mathrm{Mod}_{G(B)}, \quad M \mapsto M \otimes_{G(A)} G(B)$$

maps exact sequences of $G(A)$-modules into exact sequences of $G(B)$-modules.

A ring filtration F defines on the ring A the structure of a topological space such that the open sets are the unions of sets of type $a + F^d$ with $a \in A$. The ring filtration F is called *separated* if

$$\cap F := \bigcap_{d \in \mathbb{N}} F^d = 0,$$

which means that the topology defined by F is *Hausdorff*. The ring filtration F is called *cofinite* if the factor rings

$$A/F_A^d$$

have finite length.

1.3. Powers of ideals

Let A be a ring and $I \subseteq A$ an ideal. Then the powers of I form a ring filtration F with

$$F^d := I^d,$$

called the *I-adic filtration* of A. If A is a local ring and I is proper, then the filtration is separated by Krull's intersection theorem [Ma86], Th. 8.10. It is cofinite if and only if I is m-primary, where m denotes the maximal ideal of A. In case $I = m$, the filtration is also called *natural filtration* of A. The topology defined by the natural filtration is called the *natural topology*.

If B is an A-algebra, we will sometimes refer to the IB-adic filtration as to the I-adic filtration of B.

1.4. Symbolic powers

Let A be a ring, $P \subseteq A$ a prime ideal, and

$$F^d := P^{(d)}$$

the d-th symbolic power of P. This defines a ring filtration, which is a special case of the inverse image filtration defined below (of an ideal-adic filtration).

1.5. Weight filtrations

Let A be ring, $x_1, \ldots, x_N \in A$ elements generating a proper ideal of A and

$$q = (q_1, \ldots, q_N) \in (\mathbb{R}^+)^N$$

an N-tuple of positive real numbers. Then there is a ring filtration F with

$$F^d := \sum_{<j,q> \geq d} x^j \cdot A$$

which is called the q-adic filtration defined by $x = (x_1, \ldots, x_N)$. Here we are using multi-index notation:

$$x^j := x_1^{j_1} \cdot \ldots \cdot x_N^{j_N} \text{ and } <j, q> := \sum_{i=1}^{N} j_i q_i \text{ for } j = (j_1, \ldots, j_N).$$

Filtrations of this type are also called *weight filtrations* or *quasi-homogeneous filtrations*. A relative variant of this notion is defined as follows. Given two ring filtrations F and F' of A, the filtration F' is called q-adic over F with respect to x if

$$F'^d := \sum_{i + <j,q> \geq d} F^i \cdot x^j$$

for every d.

1.6. Inclusion, intersection, sum filtration

Let F and F' be two ring filtrations of the ring A. We say that F is *contained* in F' and write

$$F \subseteq F'$$

if $F^d \subseteq F'^d$ for every $d \in \mathbb{N}$. The *intersection* of F and F', denoted

$$F \cap F'$$

is defined by $(F \cap F')^d := F^d \cap F'^d$. The filtration $F \cap F'$ is the largest ring filtration contained in both F and F'. Note that the intersection of cofinite filtrations is cofinite:

$$L(A/F^d \cap F'^d) = L(A/F^d) + L(F^d + F'^d/F'^d) \leq L(A/F^d) + L(A/F'^d)$$

The *sum* of F and F', denoted

$$F + F'$$

is defined by $(F + F')^d := \sum_{i+j=d} F^i \cdot F'^j$. The filtration $F + F'$ is the smallest ring filtration containing both F and F'. Similarly one defines the intersection and the sum of arbitrary families of filtrations.

1.7. Sum filtrations and separatedness

The sum of separated filtrations may happen to be non-separated. Sometimes it is even rather degenerated as in the following example.

$$A := \mathbb{C}[X],$$
$$F := \text{ the } (X)\text{-adic filtration},$$
$$F' := (X + 1)\text{-adic filtration}.$$

The ideal $(F + F')^d$ contains the polynomials X^d and $(X + 1)^d$, hence the polynomials of this ideal have no common zero. By Hilbert's Nullstellensatz it is equal to A for every d.

If A is a complete local ring, the sum of two separated filtrations is again separated as is easily seen with the help of Chevalley's theorem (see 1.8 below). In the non-complete case the sum of separated filtrations in a local ring may be very well non-separated. For example (cf [Ma86], Ex. 8.10), let A be the local algebra $A := K[X, Y]_{(X,Y)}$ over some field K, take any power series $f := \sum_{i=1}^{\infty} a_i X^i \in K[[X]]$ which is transcendental over $K[X]$ and such that $a_1 \neq 0$. Define the filtrations F and G by

$$F^d := (X^d, Y - \sum_{i=1}^{d-1} a_i X^i)A \text{ and } G^d := (X^d, Y - \sum_{i=2}^{d-1} a_i X^i)A.$$

Then the intersection

$$\cap F \subseteq A \cap \left(\bigcap_{d=0}^{\infty} (F^d \cdot K[[X, Y]]) \right) = A \cap (Y - f)K[[X, Y]],$$

is zero, for, any non-zero element in the right hand side ideal would imply the existence of a non-trivial polynomial in X and Y over K with zero $Y = f$, i.e., f would be algebraic over $K[X]$. Hence F is separated. Similarly one sees that G is separated. The sum filtration is, on the contrary, far from being separated, since

$$X \in (F + G)^d \text{ for every } d.$$

However, there is a class of filtrations, called Artin-Rees filtrations, which are separated and have the property that the sum of two such filtrations is again an

Artin-Rees filtration. Artin-Rees filtrations may be considered to be the only interesting ones among the (cofinite) filtrations of a local ring. All other filtrations are degenerated in the sense that they define non-separated filtrations in the completion (as the examples given above). We will treat Artin-Rees filtrations in some detail below.

1.8. Chevalley's theorem

Let (A, m) be a complete local ring and $(F^d)_{d \in \mathbb{N}}$ a descending chain of ideals in A with

$$\bigcap_{d=0}^{\infty} F^d = (0).$$

Then for each $d \in \mathbb{N}$ there exist some $k \in \mathbb{N}$ with $F^k \subseteq m^d$.

Proof. See [Ma86], Ex. 8.7.

1.9. Separatedness of the sum in complete local rings

Let (A, m) be a complete local ring, and let F and G be separated ring filtrations of A. Then $F + G$ is separated.

Proof. The separatedness condition for F and G implies that, for every $d \in \mathbb{N}$ there is some $k(d) \in \mathbb{N}$ with $F^{k(d)} \subseteq m^d$ and $G^{k(d)} \subseteq m^d$ (by 1.8). But then $(F + G)^{2k(d)} \subseteq m^d$, i.e., $F + G$ is separated by Krull's intersection theorem ([Ma86], Th. 8.10).

1.10. Direct and inverse image filtrations

Let $f : A \to B$ be a ring homomorphism and F_B a ring filtration of B. Then the inverse images in A of the ideals F_B^d define a ring filtration of A which is called *inverse image* of F_B and denoted

$$f^*(F_B) := A \cap F_B := (f^{-1}(F_B^d))_{d \in \mathbb{N}}$$

The notation $A \cap F_B$ will be mainly used when f is the natural embedding of a subring A into B. We will refer to $A \cap F_B$ as to the filtration *induced* by F_B on A (via f). For local homomorphisms $f : (A, m) \to (B, n)$ of local rings the inverse image of a cofinite filtration is cofinite, for,

$$A/f^*(F_B)^d \subseteq B/F_B^d,$$

so the maximal ideal of $A/f^*(F_B)^d$ is nilpotent if so is the maximal ideal of B/F_B^d. On the other hand $f^*(F_B)$ cannot be separated if f is not injective.

Now let $f : A \to B$ a ring homomorphism and F_A a given ring filtration on A. Then the filtration ideals F_A^d generate ideals in B and define this way a ring filtration of B which is called *direct image* of F_A and denoted

$$f_*(F_A) := F_A \cdot B := (F_A^d \cdot B)_{d \in \mathbb{N}}$$

In case $B := A/I$ is a factor ring of A and $f : A \to A/I$ is the canonical homomorphism, $f_* F_A$ is also called the filtration *induced* by F_A on the factor ring A/I. The direct image of a separated filtration may be non-separated even if $f : A \to B$ is a local homomorphism of local rings. For example, let K be a field and

$$A := K(X, Y, Z)_{(X, Y, Z)},$$

$$F^d := (X^d, Y - Z \cdot \sum_{n=1}^{d-1} a_n X^n) A,$$

$$I = Z \cdot A,$$

where $\sum_{n=1}^{\infty} a_n X^n \in K[[X]]$ is a power series transcendental over $K[X]$. Then $F := (F^d)_{d \in \mathbb{N}}$ is a separated ring filtration. But the direct image of F under the canonical homomorphism $f : A \to A/I \cong K(X, Y)_{(X, Y)}$ is not separated: $\cap f_* F = (Y)$. Similarly, the direct image of F under the canonical embedding $A \to K[[X, Y, Z]]$ is a non-separated filtration.

The direct image of a cofinite filtration will usually not be cofinite. In the case of local homomorphisms of local rings $f : (A, m) \to (B, n)$ a necessary and sufficient condition is that the *special fibre* B/mB of f should be zero dimensional.

1.11. The filtration generated by a family

Let A be a filtered ring, $a := (a_\lambda)_{\lambda \in \Lambda}$ a family of elements of A and $(d_\lambda)_{\lambda \in \Lambda}$ a family such that

$$d_\lambda \in \mathbb{N} \cup \{\infty\} \text{ for every } \lambda \in \Lambda.$$

Define

$$(1) \qquad\qquad F^d = \sum_{(i, \lambda) \in \mathbb{N}^n \times \Lambda^n} F_A^{d - <i, d_\lambda>} \cdot a_\lambda^i$$

where the sum is taken over all non-negative integers n and, for every fixed n, over the pairs $(i, \lambda) \in \mathbb{N}^n \times \Lambda^n$. We are using the conventions,

$$a_\lambda^i := (a_{\lambda_1})^{i_1} \cdot \ldots \cdot (a_{\lambda_n})^{i_n},$$
$$d_\lambda := (d_{\lambda_1}, \ldots, d_{\lambda_n})$$
$$<i, d_\lambda> := i_1 \cdot d_{\lambda_1} + \ldots + i_n \cdot d_{\lambda_n}$$

if $i = (i_1, .., i_n)$ and $\lambda = (\lambda_1, \ldots, \lambda_n)$. The family $F = (F^d)_{d \in \mathbb{N}}$ is a ring filtration of A called the *ring filtration generated over F_A by the family $a = (a_\lambda)_{\lambda \in \Lambda}$* such that

$$(2) \qquad \qquad \operatorname{ord}(a_\lambda) \geq d_\lambda \text{ for every } \lambda \in \Lambda.$$

It is by construction the smallest filtration containing F_A such that the conditions (2) are satisfied. Note that every ring filtration F' of A with $F_A \subseteq F'$ can be obtained this way. Just let $a = (a_\lambda)_{\lambda \in \Lambda}$ be the family of all elements of A and define $d_\lambda := \operatorname{ord}_{F'} a_\lambda$. Then the filtration (1) is such that $\operatorname{ord}_F(x) \geq \operatorname{ord}_{F'}(x)$ for every $x \in A$. Thus $F'^d \subseteq F^d$. On the other hand, for $(i, \lambda) \in \mathbb{N}^n \times \Lambda^n$ the power product a_λ^i has F'-order $\geq < i, d_\lambda >$, i.e.,

$$F'^d \subseteq F^d = \sum F_A^{d - <i, d_\lambda>} a_\lambda^i \subseteq F'^d$$

hence, $F = F'$.

In case F_A is the filtration $A, 0, 0, \ldots$ one also says that F is the *filtration generated by $a = (a_\lambda)_{\lambda \in \Lambda}$* such that

$$\operatorname{ord}(a_\lambda) \geq d_\lambda \text{ for every } \lambda \in \Lambda$$

without reference to F_A.

1.12. Filtered modules, Artin-Rees filtrations

Let M be a module over the ring A. A *filtration* of M or, more precisely, an *A-module filtration* of M is a descending family $F_M = (F_M^d)_{d \in \mathbb{Z}}$ of A-submodules of M,

$$F_M^d \supseteq F_M^{d+1} \text{ for every } d.$$

A *filtered module* over a ring A is an A-module M which is equipped with an A-module filtration. If we want to stress, which filtration of the module M we have in mind, we will refer to the pair (M, F_M) as to the filtered module. We shall use the notation

$$\mathrm{G}(M) := \bigoplus_{d \in \mathbb{N}} \mathrm{G}(M)(d), \quad \text{where} \quad \mathrm{G}(M)(d) := F_M^d / F_M^{d+1}.$$

A *filtered module homomorphism* is a module homomorphism $f : M \to M'$ with M and M' filtered such that $f(F_M^d) \subseteq F_{M'}^d$. As in the case of ring filtrations, the module filtration $f^* F_{M'} = M \cap F_{M'}$ such that

$$(f^* F_M)^d := M \cap F_{M'}^d := f^{-1}(F_{M'}^d)$$

is called the *inverse image* of $F_{M'}$ under f. The definition of the *direct image filtration* $f_* F_M$ of F_M under f differs slightly from the one for ring filtrations (cf 1.10). One defines

$$(f_* F_M)^d := f(F_M^d)$$

If A is a filtered ring and the filtration F_M satisfies

$$F_A^d \cdot F_M^{d'} \subseteq F_M^{d+d'} \text{ for arbitrary } d \in \mathbb{N}, d' \in \mathbb{Z}$$

then we shall say that A and M have *compatible filtrations* and that the filtration F_M is an F_A-*filtration*. Note that, in this case $G(M)$ is a graded module over $G(A)$. If M is a filtered module and $m \in M$ an element, then

$$\operatorname{ord}(m) = \operatorname{ord}_M(m) = \operatorname{ord}_{F_M}(m) := \sup\{d \in \mathbb{N} | m \in F_M^d\}$$

is called the *order* of the element m or the F_M-*order* of m. If $\operatorname{ord}(m)$ is finite, the element

$$\operatorname{in}(m) := \operatorname{in}_{F_M}(m) := (m \bmod F_M^{\operatorname{ord}(m)+1}) \in G(M)$$

is called the *initial element* of m in $G(M)$. In case $M = A$, $\operatorname{in}(m)$ is also called *initial form* of m. We shall write

$$\operatorname{in}(m) = 0 \text{ if } \operatorname{ord}(m) = \infty.$$

Let A be a filtered ring and $I \subseteq A$ an ideal. The *initial ideal* $\operatorname{in}(I)$ of I is the ideal of $G(A)$ generated by the initial forms $\operatorname{in}(x)$ of the elements $x \in I$,

$$(1) \qquad \operatorname{in}(I) := \operatorname{in}_{F_A}(I) := \sum_{x \in I} \operatorname{in}(x) \cdot G(A).$$

Equivalently,

$$(2) \qquad \operatorname{in}(I) := \bigoplus_{d=0}^{\infty} (I \cap F_A^d + F_A^{d+1})/F_A^{d+1},$$

i.e., $\operatorname{in}(I)$ is just the kernel of the natural surjection $G(A) \to G(A/I)$ where A/I is equipped with the filtration induced by F_A. The resulting exact sequence,

$$(3) \qquad 0 \to \operatorname{in}(I) \to G(A) \to G(A/I) \to 0$$

will be sometimes called the *defining exact sequence for* $\operatorname{in}(I)$. If I and J are two ideals of the filtered ring A with $I \subseteq J$, then

$$(4) \qquad \operatorname{in}(J/I) = \operatorname{in}(J) \cdot G(A/I).$$

as is easily seen from the nine lemma applied to the commutative diagram of natural homomorphisms with exact rows and columns,

$$
\begin{array}{ccccc}
\text{in}(I) & = & \text{in}(I) & & \\
\cap & & \cap & & \\
\text{in}(J) & \subset & G(A) & \longrightarrow & G(A/J) \\
\downarrow & & \downarrow & & \| \\
\text{in}(J/I) & \subset & G(A/I) & \longrightarrow & G(A/J)
\end{array}
$$

Let A be a ring, $I \subseteq A$ an ideal, and M a filtered A-module. The filtration of M is called an *Artin-Rees filtration* with respect to I, if the following two conditions are satisfied.

1. $F_M^d \neq M$ for at least one d.

2. For every $d \in \mathbb{Z}$ there is some $k \in \mathbb{Z}$ with $IM \cap F_M^k \subseteq IF_M^d$.

It is called Artin-Rees filtration without reference to an ideal of A, if it is an Artin-Rees filtration with respect to every ideal.

Let A be a ring, M an A-module, and I an ideal of A. Then the filtration F_M of M with

$$F_M^d = I^d M$$

is called *I-adic filtration* of M. Note that the I-adic filtrations of A and M are compatible, and that the I-adic filtration of M is an Artin-Rees filtration with respect to I.

1.13. Terminology

Let A be a ring and M a filtered A-module. Then we shall frequently use the following notation.

$$
M(d) := M/F_M^{d+1}, \quad G(M)(d) := F_M^d/F_M^{d+1}, \quad \cap F_M := \bigcap_{d \in \mathbb{Z}} F_M^d
$$

$$
G(M) := \bigoplus_{d \in \mathbb{Z}} G(M)(d)
$$

$$
G(M)(\geq e) := \bigoplus_{d \geq e} G(M)(d) \quad G(M)(\leq e) := \bigoplus_{d \leq e} G(M)(d)
$$

If A is a filtered ring and M any A-module, we shall write

$$
G_{F_A}(M) := \bigoplus_{d \in \mathbb{N}} G_{F_A}(M)(d), \quad G_{F_A}(M)(d) := F_A^d M/F_A^{d+1} M.
$$

In case the filtration of A is the I-adic one where I is an ideal of A, we shall also write

$$
G_I(M) := G_{F_A}(M).
$$

1.14. Artin-Rees ring filtrations

Let A be a filtered Noetherian ring with $F_A^1 \neq A$. Then the following conditions are equivalent.

(i) F_A is an Artin-Rees filtration.

(ii) F_A is an Artin-Rees filtration with respect to F_A^1.

(iii) The topology defined by F_A on A is the F_A^1-adic topology, i.e., for every $d \in \mathbb{N}$ there is some integer k with

$$F_A^k \subseteq (F_A^1)^d$$

Proof. Suppose (iii) is satisfied, and let $I \subseteq A$ be an ideal. By the Artin-Rees lemma ([Ma86], Th. 8.5) there is a positive integer c such that

$$I \cap (F_A^1)^{d+c} \subseteq (F_A^1)^d \cdot I \text{ for every } d.$$

Fix some d and let k be such that $F_A^k \subseteq (F_A^1)^{d+c}$. Then

$$I \cap F_A^k \subseteq I \cap (F_A^1)^{d+c} \subseteq (F_A^1)^d \cdot I \subseteq F_A^d \cdot I.$$

This proves, (iii) implies (i). The implication (i)$\Rightarrow$(ii) is trivial. Suppose condition (ii) is satisfied, i.e., F_A is an Artin-Rees filtration with respect to F_A^1. We want to show that, given any $d \in \mathbb{N}$, there is some positive integer k with

$$F_A^k \subseteq (F_A^1)^d.$$

We shall prove this by induction on d. The case $d = 1$ is trivial. So assume $d > 1$. By induction hypothesis, there is some $j \in \mathbb{N}$ with

$$F_A^j \subseteq (F_A^1)^{d-1}.$$

Since F_A is an Artin-Rees filtration with respect to F_A^1, there is an integer $k > 1$ with

$$F_A^k = F_A^1 \cap F_A^k \subseteq F_A^1 \cdot F_A^j.$$

This inclusion together with the previous one implies $F_A^k \subseteq (F_A^1)^d$, as required.

1.15. Artin-Rees ring filtrations of local rings

Let (A, m) be a local ring. Then

(i) The q-adic filtration defined by x is an Artin-Rees filtration for every N-tuple $q = (q_1, \ldots, q_N)$ of positive real numbers q_i and every N-tuple $x = (x_1, \ldots, x_N)$ of elements $x_i \in m$.

(ii) If F and F' are Artin-Rees ring filtrations of A, then so are $F \cap F'$ and $F + F'$

(iii) If F is an Artin-Rees ring filtration, then so is $f_* F$ for every local homomorphism $f : A \to B$ of local rings.

(iv) If A is m-adically complete and F_A a cofinite and separated ring filtration, then F_A is an Artin-Rees filtration of A and $F_X^d := F_A^d \cdot X$ defines an Artin-Rees filtration of X for every free A-module X (possibly of infinite rank).

(v) Every Artin-Rees ring filtration F of A is separated.

(vi) An Artin-Rees ring filtration F of A is cofinite if and only if its topology is the m-adic one.

Proof. (i). Let F be the q-adic filtration defined by x. Choose some positive integer k with $\frac{1}{k} \le q_i \le k$ for every i. It is sufficient to show that (see 1.14(iii))

$$F^{d \cdot k^2} \subseteq (x)^{kd} \subseteq (F^1)^d$$

for arbitrary d where (x) denotes the ideal generated by the coordinates of x. Using multi-index notation, we have (by 1.5)

$$F^d = \sum_{<i,q> \ge d} x^i \cdot A \subseteq \sum_{|i| \cdot k \ge d} x^i \cdot A \subseteq (x)^{[d/k]}$$

which yields the first inclusion. Here $|i| = i_1 + \ldots + i_N$ for $i = (i_1, \ldots, i_N)$ and $[r]$ denotes the greatest integer $\le r$. Further

$$F^1 = \sum_{<i,q> \ge 1} x^i \cdot A \supseteq \sum_{|i|/k \ge 1} x^i \cdot A = (x)^k$$

which implies the second inclusion.

(ii). By 1.14 there is a function $k : \mathbb{N} \to \mathbb{N}$ with

$$F^{k(d)} \subseteq (F^1)^d \text{ and } F'^{k(d)} \subseteq (F'^1)^d$$

for every d. Obviously one may assume that k in monotone: $k(d) \le k(d+1)$ for every d. Then

$$(F + F')^{2k(d)} = \sum_{i+j = 2k(d)} F^i F'^j \subseteq F^{k(d)} + F'^{k(d)} \subseteq (F^1)^d + (F'^1)^d \subseteq (F^1 + F'^1)^d,$$

i.e., $F + F'$ is an Artin-Rees filtration (by 1.14(iii)). Since F is an Artin-Rees filtration with respect to the ideal $I_d := F'^{k(d)}$, there is a function $k' : \mathbb{N} \to \mathbb{N}$ with $F^{k'(d)} \cap I_d \subseteq F^{k(d)} \cdot I_d$ for every d. We may assume that $k'(d) \ge k(d)$ for every d. Then

$$(F \cap F')^{k'(d)} \subseteq F^{k'(d)} \cap I_d \subseteq F^{k(d)} \cdot I_d \subseteq (F^1)^d \cdot (F'^1)^d \subseteq (F^1 \cap F'^1)^d,$$

i.e., $F \cap F'$ is an Artin-Rees filtration (by 1.14(iii)).

(iii). Condition (iii) of 1.14 is preserved under local homomorphisms.

(iv). Let $d \in \mathbb{N}$ be arbitrary. Since F_A is cofinite, $m^{k'} \subseteq (F_A^1)^d$ for some k', and by Chevalley's theorem 1.8, $F_A^k \subseteq m^{k'}$ for some k. So F_A is an Artin-Rees filtration by 1.14(iii). Given an ideal $I \subseteq A$ and any $d \in \mathbb{N}$ there is some $k \in \mathbb{N}$ with $I \cap F_A^k \subseteq IF_A^d$. But then $IX \cap F_A^k X \subseteq IF_A^d X$ for every free A-module X.

(v). Follows from 1.14(iii) and Krull's intersection theorem ([Ma86], Th. 8.10). Note that $(F_A^1)^d \subseteq F_A^d$ and, by definition, $F_A^d \neq A$ for at least one d.

(vi). The topology defined by F_A is the F_A^1-adic topology by 1.14(iii), and $F_A^1 \subseteq m$ is a proper ideal since F_A is separated by assertion (v). So it is sufficient to show that some power of m is in F_A^1, say $m^d \subseteq F_A^1$, if and only if F_A is cofinite. But this is trivial.

1.16. Completion of filtered modules

In this section we recall, for reference purposes, part of the material in Matsumura's textbook [Ma86], chapter 8, where all this can be found in a more general form, and we fix some terminology. Our presentation is not indented as a substitute for the treatment in [Ma86], and we shall refer to the latter whenever we see a good way to do this.

Let A be a ring and M a filtered A-module. Then the module

$$M^\wedge := (M, F_M)^\wedge := \lim_{\substack{\to \\ d}} \operatorname{proj} M/F_M^{d+1}$$

$$= \left\{ (m_d + F_M^{d+1})_{d \in \mathbb{N}} \in \prod_{d \in \mathbb{N}} M/F_M^{d+1} \;\middle|\; m_d \in M, m_d - m_{d+1} \in F_M^{d+1} \right\}$$

is called the *completion* of M or the completion of the pair (M, F_M), if we want to explicitly refer to the filtration. The module M is called *complete* or F_M-complete, if the canonical mapping

$$i = i_M : M \to M^\wedge, \quad m \mapsto (m \bmod F_M^{d+1})_{d \in \mathbb{N}}$$

is an isomorphism. Let $q_d : M^\wedge \to M/F_M^{d+1}$ be the natural projection mapping the family $(m_i \bmod F_M^{i+1})_{i \in \mathbb{N}}$ to its d-th member $m_d \bmod F_M^{d+1}$ and define

$$F_{M^\wedge}^d := \operatorname{Ker}(q_{d-1} : M^\wedge \to M/F_M^d) = \{(m_i \bmod F_M^{i+1})_{i \in \mathbb{N}} \in M^\wedge | m_i \in F_M^d\}$$

Then $M^\wedge$ is a complete filtered module with respect to the filtration $F_{M^\wedge} := (F_{M^\wedge}^d)_{d \in \mathbb{N}}$. Unless stated otherwise the completion of a filtered module will be considered as filtered with respect to this filtration. By construction, each filtration module of the completion fits into a short exact sequence

$$0 \to F_{M^\wedge}^d \to M^\wedge \xrightarrow{q_{d-1}} M/F_M^d \to 0$$

which is called the *defining exact sequence for* $F_{M^\wedge}^d$.

In order to see that $M^\wedge$ is complete, note that the composition of the homomorphisms below (induced by i_M and q_d, respectively) is equal to the identity map.

$$
\begin{array}{ccccc}
M/F_M^{d+1} & \xrightarrow{\bar{\imath}_M} & M^\wedge/F_{M^\wedge}^{d+1} & \xrightarrow{\bar{q}_d} & M/F_M^{d+1} \\
m + F_M^{d+1} & \mapsto & (m + F_M^i)_{i\in\mathbb{N}} \bmod F_{M^\wedge}^{d+1} & \mapsto & m + F_M^{d+1}
\end{array}
$$

Moreover, the second homomorphism $\bar{q}_d$ is bijective, hence so is the first. Passing to the limits, one obtains two isomorphisms

$$M^\wedge \to (M^\wedge)^\wedge \to M^\wedge$$

which are mutually inverse, the first one being the canonical map $i_{M^\wedge}$. This proves, $M^\wedge$ is complete.

Given a submodule $N \subseteq M$, equip the factor module M/N with the filtration $(F_M^d + N/N)_{d\in\mathbb{Z}}$ induced by F_M and define $\mathrm{cl}(N)$ to be the kernel of the map induced by the natural homomorphism $M \to M/N$ on the completions:

$$\mathrm{cl}(N) := \mathrm{Ker}(M^\wedge \to (M/N)^\wedge) = \{(m_d \bmod F_M^{d+1}) \in M^\wedge \,|\, m_d \in N\}$$

(in particular, $\mathrm{cl}(F_M^d) = F_{M^\wedge}^d$ and $\mathrm{cl}(N) = (N, N \cap F_M)^\wedge$). The module $\mathrm{cl}(N)$ is called the *closure* of N in $(M, F_M)^\wedge$ (with respect to the topology defined by F_M). By construction, there is a short exact sequence

(1) $$0 \to \mathrm{cl}(N) \to M^\wedge \to (M/N)^\wedge \to 0$$

also called *defining exact sequence for* $\mathrm{cl}(N)$. The homomorphism on the right is easily seen to be surjective: an element of $(M/N)^\wedge$ is given by a sequence $(m_d)_{d\in\mathbb{N}}$ of elements from M such that $m_d - m_{d-1} = n_d + x_d$ where $n_d \in N$ and $x_d \in F_M^d$. But then $(m_0 + x_1 + \ldots + x_d \bmod F_M^{d+1})_{d\in\mathbb{N}}$ is an element of $M^\wedge$ and is the preimage we are looking for. The closure $\mathrm{cl}(N)$ of N can be also written as

$$\mathrm{cl}(N) := \bigcap_{d\,\in\,\mathbb{N}} (i(N) + F_{M^\wedge}^d)$$

For, if $s = (m_d \bmod F_M^{d+1})_{d\in\mathbb{N}}$ is in $\mathrm{cl}(N)$, then we may assume $m_d \in N$ for every d. Further $s - i(m_d) \in F_{M^\wedge}^d$, hence $s \in i(N) + F_{M^\wedge}^d$ for every d. This gives "$\subseteq$". The converse inclusion follows from the fact that an element from the intersection on the right is a family of elements contained in the kernel of each 'component' $M/F_M^{d+1} \to M/(N + F_M^{d+1})$ of the canonical map $M^\wedge \to (M/N)^\wedge$.

Given two module filtrations F and F' of a module M, we shall say F is a *refinement* of F' if for every d there is some $e = e(d)$ such that $F^e \subseteq F'^d$. In this situation there is a natural homomorphism

$$(M, F)^\wedge \to (M, F')^\wedge, (m_d \bmod F^{d+1}) \mapsto (m_{e(d+1)} \bmod F'^{d+1})$$

which doesn't depend upon the special choice of the function $e(d)$.

1.17. Completion of a cofinitely filtered local ring

Let (A, m) be a filtered local ring with $F_A^1 \subseteq m$ and F_A is cofinite but not necessarily separated. Then

(i) $(A, F_A)^\wedge = A^\wedge / \cap (F_A \cdot A^\wedge)$

    where $A^\wedge$ denotes m-adic completion. In particular, $(A, F_A)^\wedge$ is Noetherian, and is a local ring.

(ii) The filtration of the completion $\bar{A} := (A, F_A)^\wedge$ satisfies

$$F_{\bar{A}} = F_A \cdot \bar{A}$$

    and is an Artin-Rees filtration such that the topology defined by $F_{\bar{A}}$ is the natural topology of $\bar{A}$.

(iii) Equivalent conditions:

    a) The canonical map $A \to (A, F_A)^\wedge$ is flat.

    b) F_A is an Artin-Rees filtration.

    c) $\cap (F_A \cdot A^\wedge) = (0)$.

Proof. (i). Define

$$\bar{A} := A^\wedge / \cap (F_A \cdot A^\wedge)$$

and consider the canonical isomorphisms $A/m^k \to A^\wedge/m^k A^\wedge$ (cf [Ma86], summary of chapter 8, assertion (4)). Tensor these isomorphisms with A/F_A^d. Since F_A is cofinite, this gives an isomorphism

$$(1) \qquad\qquad A/F_A^d \to A^\wedge / F_A^d \cdot A^\wedge$$

for k large enough. Therefore,

$$(A, F_A)^\wedge = (A^\wedge, F_A \cdot A^\wedge)^\wedge = (\bar{A}, F_A \cdot \bar{A})^\wedge$$

The ring $\bar{A}$ is complete with respect to the natural topology (see [Ma86], Th. 8.1(ii)). As a separated and cofinite ring filtration in a complete local ring, $F_A \cdot \bar{A}$ is an Artin-Rees filtration (by 1.15(iv)). Thus, the completion on the right is the usual completion with respect to the natural topology (by 1.15(vi)). Since $\bar{A}$ is already complete in this topology,

$$(\bar{A}, F_A \cdot \bar{A})^\wedge = \bar{A}.$$

This proves (i).

(ii). The filtration $F_{\bar{A}}$ of $\bar{A} = (A, F_A)^\wedge$ is given by the defining exact sequences (see 1.16)

$$0 \to F_{\bar{A}}^d \to \bar{A} \xrightarrow{\bar{q}_{d-1}} A/F_A^d \to 0.$$

Tensor with A/F_A^d over A to obtain an exact sequence

$$F_{\bar{A}}^d \to \bar{A}/F_A^d \bar{A} \xrightarrow{\bar{q}_{d-1}} A/F_A^d \to 0,$$

which shows that $F_A^d \bar{A} \subseteq F_{\bar{A}}^d$ and $F_{\bar{A}}^d/F_A^d \bar{A} = \mathrm{Ker}(\bar{q}_{d-1})$. Hence the identity $F_{\bar{A}} = F_A \cdot \bar{A}$ of (ii) will follow if we can prove $\bar{q}_{d-1} := q_{d-1} \otimes_A A/F_A^d$ is injective. Consider the decomposition of the identity map,

$$A/F_A^d \xrightarrow{\bar{i}} \bar{A}/F_A^d \bar{A} (= A^\wedge/F_A^d \cdot A^\wedge) \xrightarrow{\bar{q}_{d-1}} A/F_A^d$$

where $\bar{i}$ is induced by the natural embedding $i_{\bar{A}} : A \to \bar{A}$. From the proof of assertion (i) we know that $\bar{i}$ is an isomorphism (see (1)), hence so is $\bar{q}_{d-1}$. This proves $F_{\bar{A}} = F_A \cdot \bar{A}$. We have already seen in the proof of (i) that $F_A \cdot \bar{A}$ is an Artin-Rees filtration defining the natural topology.

(iii). a) $\Rightarrow$ c). As above let

$$\bar{A} := A^\wedge/I, \quad I := \bigcap_{d \in \mathbb{N}} F_A^d \cdot A^\wedge,$$

be the completion of (A, F_A). Assume, c) is wrong, i.e, the ideal I is non-zero. By Krull's intersection theorem, I cannot be contained in every power of $m \cdot A^\wedge$, say $I \not\subseteq m^k \cdot A^\wedge$. But then the homomorphism

$$(A \to \bar{A}) \otimes_A A/m^k = \begin{array}{ccc} A/m^k & \to & \bar{A}/m^k \bar{A} \\ \| & & \| \\ A^\wedge/m^k A^\wedge & & A^\wedge/(m^k A^\wedge + I) \end{array}$$

is not injective, i.e., $\bar{A}$ is not flat over A.

c) $\Rightarrow$ b). It is sufficient to show that condition 1.14(iii) is satisfied. Since F_A is cofinite, for every given $d \in \mathbb{N}$ there is some $d' \in \mathbb{N}$ with

$$m^{d'} \subseteq (F_A^1)^d.$$

Condition c) implies (by Chevalley's theorem 1.8)

$$F_A^k \subseteq F_A^k A^\wedge \cap A \subseteq m^{d'} A^\wedge \cap A = m^{d'}$$

for k large enough, hence $F_A^k \subseteq (F_A^1)^d$ as required.

b) $\Rightarrow$ a). Condition b) implies by 1.14(iii), that the topology defined by F_A is the F_A^1-adic topology. But the F_A^1-adic completion of A is flat over A ([Ma86], Th. 8.8).

1.18. Closure and Artin-Rees filtrations

Let M be a filtered A-module and $I \subseteq A$ an ideal. Then

$$IM^\wedge \subseteq \mathrm{cl}(IM)$$

Suppose the following two conditions are satisfied.

1. I is finitely generated.
2. F_M is an Artin-Rees filtration with respect to I.

Then equality holds above, $\mathrm{cl}(IM) = IM^\wedge$, and the canonical homomorphism

$$M^\wedge / I \cdot M^\wedge \to (M/IM)^\wedge$$

is an isomorphism.

Proof. From the defining exact sequence for $\mathrm{cl}(IM)$ (see 1.16),

$$0 \to \mathrm{cl}(IM) \to M^\wedge \to (M/IM)^\wedge \to 0,$$

we see that there is a natural isomorphism $M^\wedge / \mathrm{cl}(IM) \cong (M/IM)^\wedge$. So the claimed inclusion is just a consequence of the fact that the module $(M/IM)^\wedge$ is annihilated by I. Now assume that conditions 1 and 2 above are satisfied. We have to show that the kernel of the canonical map

$$f : M^\wedge \to (M/IM)^\wedge$$

(which is equal to $\mathrm{cl}(IM)$ by definition) is contained in $IM^\wedge$. Write

$$I = (a_1, \dots, a_r)A.$$

Since F_M is an Artin-Rees filtration with respect to I, there exists a function $k : \mathbb{N} \to \mathbb{N}$ such that $IM \cap F_M^{k(d)} \subseteq IF_M^{d+1}$ for every d. We may assume that $k(d)$ is monotone: $k(d) \le k(d')$ for $d \le d'$. Let $s = (m_d \bmod F_M^{d+1}) \in \mathrm{Ker}(f)$. Then we may assume $m_d \in IM$ for every d. Since $m_d - m_{d+1} \in F_M^{d+1}$ for every d,

$$n_d := m_{k(d)} - m_{k(d+1)} \in IM \cap F_M^{k(d)} \subseteq IF_M^{d+1}.$$

So the elements n_d can be written

$$n_d = \sum_{i=1}^{r} a_i m_d^i, \quad m_d^i \in F_M^{d+1}$$

Let $m^i := \sum_{d=0}^{\infty} m_d^i$, which is a well-defined element of $M^\wedge$. Moreover,

$$\sum_{i=1}^{r} a_i m^i = \lim_{d \to \infty} \sum_{i=1}^{r} a_i \sum_{j=0}^{d} m_j^i$$

$$= \lim_{d \to \infty} \sum_{j=0}^{d} n_j = m_{k(0)} - \lim_{d \to \infty} m_{k(d)} = m_{k(0)} - s.$$

Therefore $s - m_{k(0)} \in IM^\wedge$. Since $m_{k(0)}$ is in IM, the element s must be in $IM^\wedge$, as required.

2. Basic lemmas

Remark

This chapter contains technical results which are frequently used in the sequel. Mostly these are versions of well-known theorems reformulated according to our needs. In first reading one might skip them and return to this chapter when the results are needed.

2.1. Graded Nakayama lemma

Let G be a graded ring, M a graded G-module, $N \subseteq M$ a graded G-submodule, and $I \subseteq G$ a homogeneous ideal such that

$$M = N + IM.$$

Assume that one of the following conditions is satisfied.

1. The degree zero part $I(0)$ of I is contained in the radical of $G(0)$ and the homogeneous parts of M/N are finitely generated over $G(0)$.
2. Each $(M/N)(d)$ is annihilated by some power of $I(0)$.

Then $M = N$.

Proof. First assume that the second condition is satisfied. Given any $d \in \mathbb{N}$, choose $r \in \mathbb{N}$ such that $I(0)^{r+1}$ annihilates $(M/N)(i)$ for every $i \leq d$, i.e.,

$$I(0)^{r+1} M(\leq d) \subseteq N$$

Then, since $I \subseteq I(0) + G^{+}$, hence $I^n \subseteq \sum_{i=0}^{n} I(0)^i \cdot (G^+)^{n-i}$,

$$
\begin{aligned}
I^n M &= I^n M(\leq d) + I^n M(> d) \\
&\subseteq \sum_{i=0}^{r} I(0)^i \cdot (G^+)^{n-i} M(\leq d) + N + M(> d) \\
&\subseteq (G^+)^{n-r} M(\leq d) + N + M(> d).
\end{aligned}
$$

Since M is zero in 'very negative' degrees,

$$I^n M \subseteq N + M(> d)$$

for sufficiently large n, say $n = n(d)$. Iterating the identity $M = N + IM$ one obtains

$$M = N + I^n M \subseteq N + M(> d).$$

In particular, $M(d) \subseteq N(d)$, i.e., $M(d) = N(d)$. Since d was arbitrary, $M = N$ as required.

Now suppose the first condition of the lemma holds. Then $I \subseteq \operatorname{rad} G(0) + G^+$, hence

$$M = N + (\operatorname{rad} G(0))M + G^+ M.$$

Since the degree zero part of G^+ is zero (hence nilpotent), the first part of the proof gives $M = N + (\operatorname{rad} G(0)) \cdot M$, i.e.,

$$M(d) = N(d) + \operatorname{rad} G(0) \cdot M(d).$$

The usual Nakayama lemma ([Ma86], Corollary to Th. 2.2) implies $M(d) = N(d)$. Since d was arbitrary, $M = N$ as claimed.

2.2. Flatness of graded modules

Let G be a graded ring and M a graded G-module. Then the following conditions are equivalent.

(i) M is flat over G.

(ii) Given homogeneous elements $m_i \in M(d_i)$, $g_i \in G(d - d_i)$, $i = 1,\ldots,r$ with

$$\sum_{i=1}^{r} g_i m_i = 0,$$

there are homogeneous elements $m'_j \in M(d'_j)$, $g_{ij} \in G(d_i - d'_j)$, $i = 1,\ldots,r$, $j = 1,\ldots,s$ such that the following two conditions are satisfied.

   a) $m_i = \sum_{j=1}^{s} g_{ij} m'_j$ for $i = 1, \ldots, r$.
   b) $0 = \sum_{i=1}^{r} g_i g_{ij}$ for $j = 1, \ldots, s$.

Proof(see [Ma86], proof of Th. 7.6). (i)$\Rightarrow$(ii). Consider the graded G-module homomorphisms

$$\phi : \quad S := \bigoplus_{i=1}^{r} G[-d + d_i] \to G$$

$$\phi_M : \quad S_M := \bigoplus_{i=1}^{r} M[-d + d_i] \to M$$

defined by $(x_1, \ldots, x_r) \mapsto g_1 x_1 + \ldots + g_r x_r$. Note that, up to isomorphism, $\phi_M = \phi \otimes_G M$. Apply the functor $\otimes_G M$ to the exact sequence

$$0 \to \operatorname{Ker}(\phi) \to S \xrightarrow{\phi} G.$$

Since, by assumption (i), M is flat over G, the resulting sequence,

$$0 \to \operatorname{Ker}(\phi) \otimes_G M \to S_M \xrightarrow{\phi_M} M$$

is also exact. The element $(m_1, \ldots, m_r) \in S_M$ is homogeneous of degree d, and is in the kernel of ϕ_M, hence equal to the image of some degree d element α of $\mathrm{Ker}(\phi) \otimes_G M$. Write $\alpha = \sum_{j=1}^{s}(g_{1j}, \ldots, g_{rj}) \otimes m_j'$ with homogeneous elements $(g_{1j}, \ldots, g_{rj}) \in \mathrm{Ker}(\phi)$, $m_j' \in M(d_j')$, $g_{ij} \in G(d_i - d_j')$. Then

$$(m_1, \ldots, m_r) = \sum_{j=1}^{s}(g_{1j} \cdot m_j', \ldots, g_{rj} \cdot m_j')$$

hence condition a) of assertion (ii) is satisfied. Since condition b) is just a reformulation of the fact that $(g_{1j}, \ldots, g_{rj})$ is in $\mathrm{Ker}(\phi)$, the proof of the implication is complete.

(ii)$\Rightarrow$(i). It will be sufficient to show, the canonical homomorphism

$$H \otimes_G M \to M, h \otimes m \mapsto hm,$$

is injective for every finitely generated homogeneous ideal H of G (see [H-O82], Prop. (3.1)). Write

$$H = Gg_1 + \ldots + Gg_r$$

with $g_i \in G(d - d_i)$ and let $\alpha = \sum_{i=1}^{r} g_i \otimes m_i \in H \otimes_G M$, $m_i \in M(d_i)$, be a homogeneous element of degree d with image zero in M. We have to show that it is already zero itself. Since α is mapped to zero, $\sum_{i=1}^{r} g_i m_i = 0$. So there are elements $m_j' \in M(d_j')$, $g_{ij} \in G(d_i - d_j')$ satisfying conditions a) and b) of (ii). But then

$$\alpha = \sum_{i=1}^{r} g_i \otimes m_i = \sum_{j=1}^{s}\sum_{i=1}^{r} g_i \otimes g_{ij} m_j' = \sum_{j=1}^{s}\sum_{i=1}^{r} g_i g_{ij} \otimes m_j' = 0,$$

as required.

2.3. Local flatness criterion, graded case

Let G be a graded ring, M a graded G-module, and F_G, F_M module filtrations by homogeneous submodules of G and M, respectively. Assume that for every $d \in \mathbb{Z}$ the following conditions are satisfied.

1. M/F_M^d is a flat module over G/F_G^d (in particular, F_G^d annihilates M/F_M^d).
2. There is some $e \in \mathbb{Z}$ such that the canonical homomorphisms $G \to G/F_G^e$ and $M \to M/F_M^e$ are injective in degrees $\leq d$ (i.e., F_G^e and F_M^e are generated by homogeneous elements of degree $> d$).

Then M is a flat G-module.

Proof. It will be sufficient to show that the conditions of 2.2(ii) are satisfied. Let $m_i \in M(d_i)$, $g_i \in G(d - d_i)$, $i = 1, \ldots, r$ be homogeneous elements with $\sum_{i=1}^{r} g_i m_i = 0$ where d is such that $d_i \leq d$ for every i. Choose some $e \in \mathbb{Z}$ such that the canonical maps $\rho : G \to G/F_G^e$ and $\sigma : M \to M/F_M^e$ are injective in degrees $\leq d$. Then $\sum_{i=1}^r \rho(g_i)\sigma(m_i) = 0$. Since M/F_M^e is flat over G/F_G^e, there are homogeneous elements $m_j' \in M(d_j')$, $g_{ij} \in G(d_i - d_j')$ with

$$\sigma(m_i) = \sum_{j=1}^s \rho(g_{ij})\sigma(m_j') \text{ for } i = 1, \ldots, r \text{ and}$$

$$0 = \sum_{i=1}^r \rho(g_i g_{ij}) \text{ for } j = 1, \ldots, s.$$

These identities mean that the elements

$$m_i - \sum_{j=1}^s g_{ij} m_j' \quad \text{and} \quad \sum_{i=1}^r g_i g_{ij}$$

are mapped to zero under σ and ρ, respectively. But the elements have degree

$$d_i \leq d \quad , \text{respectively}, \quad d - d_j' \leq d$$

whereas σ and ρ are injective in degrees $\leq d$, so the elements themselves are already zero,

$$m_i = \sum_{j=1}^s g_{ij} m_j' \quad \text{and} \quad 0 = \sum_{i=1}^r g_i g_{ij}.$$

Thus M satisfies the conditions of the flatness criterion for graded modules, 2.2.

2.4. Refinement of a topology

Let X be a set equipped with a topology T. A *refinement* of T is a topology T' of X such that every T-open set is also T'-open. Usually the notions of *stronger* and *weaker* topologies are applied in this context, but there seems to be no generally accepted convention which one to call stronger and which one weaker. Therefore, we have decided to use the term "refinement".

A variant of the lemma that follows can be found in the Elements of Algebraic Geometry [G-D71] (EGA $0_I(6.6.21)$ and $0_I(7.1.14)$).

2.5. Filtered Grothendieck lemma

Let A be a ring, $I \subseteq A$ an ideal, and $f : M \to N$ a homomorphism of filtered modules over A. Assume the following conditions are satisfied.

1. The I-adic topologies on M and N are refinements of the topologies defined by F_M and F_N, respectively.
2. F_N is separated.

Then

(i) If $f \otimes_A A/I : M/IM \to N/IN$ is surjective and M is complete, then $f : M \to N$ is surjective, too.

(ii) If $f \otimes_A A/I : M/IM \to N/IN$ is bijective, M is complete, and N is flat over A, then $f : M \to N$ is bijective, too.

Note that, according to our conventions, the requirement that M is complete (without explicit reference to a filtration) means that M is F_M-complete.

Proof. (i). Consider the commutative diagram,

$$
\begin{array}{ccc}
M & \xrightarrow{f} & N \\
\downarrow & & \downarrow \\
M/IM & \xrightarrow{f_0} & N/IN
\end{array}
$$

where the vertical maps are the canonical surjections and $f_0 := f \otimes_A A/I$. Let y be an element of N. Since f_0 is surjective, there is an element $x \in M$ with $y - f(x) \in IN$. This implies, since f is A-linear, that for every $y \in I^k N$ there is some $x \in I^k M$ with $y - f(x) \in I^{k+1} N$. Now fix some $y \in N$. From the above we see that there is a sequence of elements $x_k \in I^k M$ with $y - \sum_{k=0}^{n} f(x_k) \in I^{n+1} N$ for $n = 0,1,2,\dots$ The series $\sum_{k=0}^{\infty} x_k$ converges I-adically hence in the topology defined by F_M. So $x := \sum_{k=0}^{\infty} x_k$ is a well-defined element of the complete module M. For every $d \in \mathbb{N}$ there is some $n = n(d) \in \mathbb{N}$ such that $x - \sum_{k=0}^{n} x_k \in F_M^d$ hence $f(x) - \sum_{k=0}^{n} f(x_k) \in F_N^d$. We may assume $n = n(d)$ is so large that the difference $y - \sum_{k=0}^{n} f(x_k) (\in I^{n+1} N)$ is also in F_N^d. Then $f(x) - y \in F_N^d$. Since d is arbitrary, $f(x) = y$. We have proved that f is surjective.

(ii). It will be sufficient to show that f is injective. Define

$$
f_d := f \otimes_A A/I^{d+1} : M/I^{d+1}M \to N/I^{d+1}N
$$

and consider the exact sequences

$$
0 \to \mathrm{Ker}(f_d) \to M/I^{d+1}M \xrightarrow{f_d} N/I^{d+1}N \to 0.
$$

Note that since f_0 is surjective, $N = \mathrm{Im}(f) + IN$ hence $N = \mathrm{Im}(f) + I^{d+1}N$ so that f_d is surjective for every d. By assumption, the module $N/I^{d+1}N$ is flat over A/I^{d+1}. Thus, tensoring with A/I, we get again a short exact sequence. Since the map $f_d \otimes A/I = f_0$ is bijective, $\mathrm{Ker}(f_d) = I \cdot \mathrm{Ker}(f_d)$ hence $\mathrm{Ker}(f_d) = 0$ (since I^{d+1} annihilates $\mathrm{Ker}(f_d)$). We have proved, the homomorphisms f_d are bijective. So there is a commutative diagram

$$
\begin{array}{ccc}
\lim\mathrm{proj}\, M/I^{d+1}M & \to & \lim\mathrm{proj}\, N/I^{d+1}N \\
\uparrow & & \uparrow \\
M & \xrightarrow{f} & N
\end{array}
$$

with the upper horizontal homomorphism being an isomorphism. To prove that f is injective, it will be sufficient to show that the vertical homomorphism on the left is injective. Compose the latter with the natural homomorphism $\lim \text{proj}\, M/I^{d+1}M \to (M, F_M)^\wedge$ (which exists by assumption 1 and 1.16). To prove the claim, it will be sufficient to show that this composition,

$$M \to \lim \text{proj}\, M/I^{d+1} \to (M, F_M)^\wedge$$

is injective. But this is just the natural map of M into its completion which is an isomorphism, since M is complete.

2.6. Free generators of the associated graded module

Let A be filtered ring and M a filtered A-module such that F_M is an F_A-filtration. Moreover, let $(m_\lambda)_{\lambda \in \Lambda}$ be a family of elements from M and define

$$\mathrm{d}(\lambda) := \mathrm{ord}_{F_M}(m_\lambda)$$

Then the following conditions are equivalent.

(i) The family $\mathrm{in}(m) := (\mathrm{in}(m_\lambda))_{\lambda \in \Lambda}$ is a free generating system of $G(M)$ over $G(A)$.

(ii) The following two conditions are satisfied.

    (a) $M \subseteq \sum_{\lambda \in \Lambda} Am_\lambda + F_M^d$ for every $d \in \mathbb{Z}$.

    (b) $\sum_{\lambda \in \Lambda} a_\lambda m_\lambda \in F_M^d$ with $a_\lambda \in A$ for every λ implies that $a_\lambda \in F_A^{d-\mathrm{d}(\lambda)}$ for every λ.

(iii) The homomorphism

$$f_d : \bigoplus_{\lambda \in \Lambda} (A/F_A^{d-\mathrm{d}(\lambda)}) \to M/F_M^d$$

induced by $(a_\lambda)_{\lambda \in \Lambda} \mapsto \sum_{\lambda \in \Lambda} a_\lambda m_\lambda$ is an isomorphism for every $d \in \mathbb{Z}$.

Proof. (i)$\Rightarrow$(ii). Let $x \in M$ be an element with finite order d. Then $\mathrm{in}(x) \in G(M)$ is a linear combination of the initial forms $\mathrm{in}(m_\lambda)$, so that

$$x \in \sum_{\lambda \in \Lambda} F_A^{d-\mathrm{d}(\lambda)} \cdot m_\lambda + F_M^{d+1}.$$

This shows $F_M^d \subseteq \sum_{\lambda \in \Lambda} F_A^{d-\mathrm{d}(\lambda)} \cdot m_\lambda + F_M^{d+1}$. Iterating this inclusion, condition (a) of (ii) is easily derived. Next assume, condition (b) is not satisfied for some expression $\sum a_\lambda m_\lambda \in F_M^d$. Then the inequality $\mathrm{ord}(a_\lambda) \geq d - \mathrm{d}(\lambda)$ doesn't hold for at least one $\lambda \in \Lambda$, i.e.,

$$i := \min\{\mathrm{ord}(a_\lambda) + \mathrm{d}(\lambda) | \lambda \in \Lambda\}$$

is strictly less than d. In particular, $F_M^d \subseteq F_M^{i+1}$, hence

$$0 = (\sum a_\lambda m_\lambda \bmod F_M^{i+1}) = \sum (a_\lambda \bmod F_A^{i+1-d(\lambda)}) \cdot \text{in}(m_\lambda).$$

Since the elements $\text{in}(m_\lambda)$ form a free basis, $a_\lambda \in F_A^{i+1-d(\lambda)}$ for every $\lambda \in \Lambda$. But this contradicts the definition of i.

(ii)$\Rightarrow$(iii). Since $m_\lambda \in F_M^{d(\lambda)}$ for every $\lambda \in \Lambda$, the map f_d is well-defined. It is surjective by property (a) of (ii) and injective by property (b).

(iii)$\Rightarrow$(i). Consider the following commutative diagram with exact columns (see 1.13 for terminology).

$$
\begin{array}{ccc}
0 & & 0 \\
\downarrow & & \downarrow \\
\oplus_{\lambda \in \Lambda}\, G(A)(d - d(\lambda)) & \longrightarrow & G(M)(d) \\
\downarrow & & \downarrow \\
\oplus_{\lambda \in \Lambda}\, A(d - d(\lambda)) & \xrightarrow{f_d} & M(d) \\
\downarrow & & \downarrow \\
\oplus_{\lambda \in \Lambda}\, A(d - d(\lambda) - 1) & \xrightarrow{f_{d-1}} & M(d - 1) \\
\downarrow & & \downarrow \\
0 & & 0
\end{array}
$$

Here the two lower horizontal homomorphisms f_d and f_{d-1} are bijective by assumption (iii), hence so is the upper one. Taking the direct sum over $d \in \mathbb{Z}$, one obtains an isomorphism over $G(A)$,

$$\mathop{\oplus}_{\lambda \in \Lambda} G(A)[-d(\lambda)] \to G(M), \quad (a_\lambda)_{\lambda \in \Lambda} \mapsto \sum_{\lambda \in \Lambda} a_\lambda \cdot \text{in}(m_\lambda)$$

which shows that $\text{in}(m)$ is a free generating system of $G(M)$ over $G(A)$.

2.7. Lifting free generators to a flat module

Let A be a ring, M a flat A-module, $I \subseteq A$ an ideal, and $(m_\lambda)_{\lambda \in \Lambda}$ a family of elements from M. Assume that the following conditions are satisfied.

1. M/IM is A/I-free and the residue classes

$$\bar{m}_\lambda := (m_\lambda \bmod IM)$$

are part of a free generating system of M/IM over A/I.

2. The I-adic topology of A is separated,

$$\bigcap_{d=0}^{\infty} I^d = (0).$$

Then

(i) The elements m_λ are linearly independent over A.

(ii) $\mathrm{Tor}_1^A(A/J, M/\sum_{\lambda \in \Lambda} A m_\lambda) = 0$ for every ideal $J \subseteq A$ such that A/J is I-adically separated.

Proof. (i). We may assume that the elements $\bar{m}_\lambda$ form a free generating system of M/IM over A/I. Consider the homomorphism

$$f : X := \bigoplus_{\lambda \in \Lambda} A \to M, (a_\lambda)_{\lambda \in \Lambda} \mapsto \sum_{\lambda \in \Lambda} a_\lambda m_\lambda.$$

By assumption, $f \otimes A/I$ is an isomorphism. In particular, $M = \mathrm{Im}(f) + IM$, hence $M = \mathrm{Im}(f) + I^n M$ for arbitrary $n \in \mathbb{N}$, and the homomorphism

$$f_n := f \otimes_A A/I^n : X/I^n X \to M/I^n M$$

is surjective for every n. Let K_n denote its kernel. Since $M/I^n M$ is flat over A/I^n, tensoring f_n with A/I gives an exact sequence

$$0 \to K_n \otimes A/I \to X/IX \xrightarrow{f \otimes A/I} M/IM \to 0,$$

and the fact that $f \otimes A/I$ is bijective, implies $K_n = IK_n$, hence $K_n = I^n K_n = 0$. We have proved, f_n is an isomorphism for every n. Taking the inverse limit, we get an isomorphism $f^\wedge : X^\wedge \to M^\wedge$ of the I-adic completions, which can be inserted into a commutative diagram as follows.

$$\begin{array}{ccc} X & \xrightarrow{f} & M \\ \downarrow & & \downarrow \\ X^\wedge & \xrightarrow{f^\wedge} & M^\wedge \end{array}$$

Here the vertical homomorphisms are the natural ones. Since A is I-adically separated, the free module X maps injectively into its I-adic completion $X^\wedge$. So bijectivity of $f^\wedge$ implies that f is at least injective. But this is the claim of (i).

(ii). Since the elements $m_\lambda \in M$ are linearly independent over A, there is a short exact sequence

$$(1) \qquad 0 \to X \to M \to M/(\sum_{\lambda \in \Lambda} A m_\lambda) \to 0.$$

Here, as above, $X := \bigoplus_{\lambda \in \Lambda} A$. Let $J \subseteq A$ be a proper ideal such that A/J is I-adically separated. Define

$$\bar{A} := A/J, \quad \bar{M} := M/JM, \quad \bar{I} := I \cdot \bar{A}.$$

Then the assumptions of the lemma are also satisfied with A, M, I replaced by $\bar{A}$, $\bar{M}$, $\bar{I}$, respectively. In particular, the residue classes modulo JM of the elements m_λ are linearly independent over A/J. This gives an exact sequence

$$0 \to X/JX \to M/JM \to M/(JM + \sum_{\lambda \in \Lambda} Am_\lambda) \to 0.$$

which is just the sequence (1) tensored with A/J. Therefore,

$$\operatorname{Tor}_1^A(A/J, M/(\sum_{\lambda \in \Lambda} Am_\lambda)) = 0$$

as required.

2.8. Length of a tensor product

Let $f : A \to B$ be a ring homomorphism, M an A-module of finite length, and N a B-module. Then,

(i) $L_B(M \otimes_A N) \leq \sum_{m \in \Omega(A)} L_{A_m}(M_m) L_B(N/mN)$

(ii) If N is A-flat, equality holds in (i).

(iii) If A is Artinian, N has finite length over B, and

$$L_B(N) = \sum_{m \in \Omega(A)} L(A_m) L_B(N/mN)$$

(i.e., equality holds in (i) for $M = A$), then N is A-flat.

These assertions are more or less generally known (and elementary). But we do not know a good reference for them and therefore decided to give the full proofs.

Proof. Let $t := L_A(M)$ and consider a maximal chain of submodules

$$0 = M_0 \subset M_1 \subset \ldots \subset M_t = M,$$

i.e., such that $L_A(M_{j+1}/M_j) = 1$ for every j. Then $M_{j+1}/M_j \cong A/m_j$ with m_j a maximal ideal of A. From the exact sequences

$$0 \to M_j \to M_{j+1} \to A/m_j \to 0,$$

we get, applying the functor $\otimes_A N$,

$$L_B(M_{j+1} \otimes_A N) \leq L_B(M_j \otimes_A N) + L_B(N/m_j N).$$

For every maximal ideal $m \in \Omega(A)$ define

$$\nu(m) := \#\{j : m_j = m\} = L_{A_m}(M_m).$$

Then

$$L_B(M \otimes_A N) \leq \sum_{j=0}^{t-1} L_B(N/m_j N) = \sum_{m \in \Omega(A)} \nu(m) \cdot L_B(N/mN)$$

$$= \sum_{m \in \Omega(A)} L_{A_m}(M_m) \cdot L_B(N/mN).$$

This proves assertion (i). Note that, if N is A-flat, equality holds in all the estimations above, i.e., assertion (ii) is true. Now suppose, A is Artinian, N has finite length over B, and equality holds in (ii) for $M = A$, i.e.,

$$(2) \qquad L_B(N) = \sum_{m \in \Omega(A)} L(A_m) L_B(N/mN).$$

We want to show that N is A-flat. As a first step we reduce the proof of this assertion to the local case. Let $n \in \Omega(B)$ and $m := f^{-1}(n)$. Note that the ideal m is maximal in A since A is Artinian. Consider the local homomorphism $f_n : A_m \to B_n$ induced by f. It will be sufficient to show f_n is flat for every n. Assertion (i) applied to f_n gives the inequality

$$(3) \qquad L_{B_n}(N_n) \leq L(A_m) L_{B_n}(N_n/mN_n).$$

Taking the sum over all maximal ideals n of B we get

$$\sum_{n \in \Omega(B)} L_{B_n}(N_n) \leq \sum_{m \in \Omega(A)} L(A_m) \sum_{\substack{n \in \Omega(B) \\ f^{-1}(n)=m}} L_{B_n}(N_n/mN_n)$$

Since B is flat over itself, the left hand side of this inequality equals $L_B(N \otimes_B B) = L_B(N)$ (in view of assertion (ii)). Similarly, the inner sum on the right is equal to $L_B(N/mN)$. So the inequality reads

$$L_B(N) \leq \sum_{m \in \Omega(A)} L(A_m) L_B(N/mN).$$

In other words, if we take the sum over all $n \in \Omega(B)$ of the inequalities (3), we get identity (2). But this is only possible if equality holds in (3) for every $n \in \Omega(B)$. We have proved that the map f_n and the module N_n satisfy the conditions of assertion (iii), i.e., the proof of (iii) is reduced to the case that f is a local homomorphism $(A, m) \to (B, n)$ of local rings and N is a B-module such that

$$L_B(N) = L(A) \cdot L_B(N/mN).$$

We have to show that N is B-flat. Consider the canonical surjection

$$G(A) \otimes_A N \to G(N),$$

where A and N are equipped with the m-adic filtrations. By the local flatness criterion ([Ma86], Th. 22.3(4')), it is sufficient to show that this map is bijective. Since the modules on both sides have finite length, it will be enough to prove

$$L_B(G(A) \otimes_A N) \leq L_B(G(N)).$$

The right hand side length is equal to $L_B(N)$ and the length on the left can be estimated as follows,

$$L_B(G(A) \otimes_A N) \leq L_A(G(A)) \cdot L_B(N/mN) = L(A) \cdot L_B(N/mN).$$

We have used here assertion (i) again. The product on the right is, by assumption, equal to $L_B(N)$, which gives the required inequality.

2.9. Refinement and completion

Let M be an A-module and F, F' two module filtrations of M such that

1. (M, F) is complete.

2. $F' \subseteq F$.

3. Each submodule F'^d is closed in the topology defined by the filtration F.

Then the natural homomorphism $f : (M, F')^\wedge \to (M, F)^\wedge$ is bijective.

Proof. Since (M, F) is complete, the composition

$$M \xrightarrow{i'} (M, F')^\wedge \xrightarrow{f} (M, F)^\wedge$$

of f with the natural homomorphism $i' : M \to (M, F')^\wedge$ is an isomorphism. So f is surjective. Let $s := (m_d \bmod F'^{d+1})_{d \in \mathbb{N}}$ be in the kernel of f. Then, $m_d \to 0$ in the topology defined by F. Let N be an arbitrary submodule of M, and define $\bar{m}_d := (m_d \bmod N)$. Then $\bar{m}_d \to 0$ in the topology induced by F on M/N. Now assume s is a non-zero element of $(M, F')^\wedge$. This means, there is some positive integer i such that none of the elements m_d is in F'^i. Let $N := F'^i$. Then all the elements of the sequence $(\bar{m}_d)_{d \in \mathbb{N}}$ are non-zero. Moreover, this sequence is stationary (since $s \in (M, F')^\wedge$). But then the limit of this sequence cannot be zero. This contradiction shows that f must be injective.

2.10. Tjurina's flatness criterion

Let $A \to B$ be a local homomorphism of local rings, M a finitely generated B-module that is flat over A, and $N = Bn_1 + \ldots + Bn_r$ a B-submodule of M. Write

$$\bar{n}_i := (n_i \bmod mM)$$

for the residue class of n_i in M/mM where m denotes the maximal ideal of A. Then the following assertions are equivalent.

(i) M/N is flat over A.

(ii) $mM \cap N = mN$.

(iii) Every B-linear relation in M/mM of the given generators,

$$b_1 \bar{n}_1 + \ldots + b_r \bar{n}_r = 0,$$

can be lifted to a relation in M, i.e., there are elements $b_i' \in B$ with $b_i' \equiv b_i \bmod mB$ for every i such that

$$b_1' n_1 + \ldots + b_r' n_r = 0.$$

Note that it is sufficient to check condition (iii) for a generating system of the syzygy module $R(\bar{m}) := \{ \bar{b} \in (B/mB)^r \mid < \bar{b}, \bar{n} >= 0 \}$.

Proof. (i)⇔(ii). This is essentially Theorem 22.5 of [Ma86].

(ii)⇒(iii). Write

$$n := (n_1, \ldots, n_r)$$
$$b := (b_1, \ldots, b_r)$$
$$< b, n >:= b_1 n_1 + \ldots + b_r n_r$$

Then the B-linear relation in M/mM given in (iii) implies that $< b, n > \in mM$, hence $< b, n > \in mM \cap N = mN$ by assumption (ii). There is an r-tuple $x = (x_1, \ldots, x_r)$ of elements from mB such that $< b, n >=< x, n >$ or, equivalently, $< b - x, n >= 0$. But then, the elements $b_i' := b_i - x_i$ have the property required in (iii).

(iii)⇒(ii). The inclusion "⊇" is trivial. So let $\alpha \in mM \cap N$ be an element from the left hand side of the identity to be proved. Then there is an r-tuple $b = (b_1, \ldots, b_r)$ of elements from B such that

$$\alpha =< b, n > \in mM.$$

This corresponds to a B-linear relation in M/mM which, by assumption can be lifted to M, i.e., there is an r-tuple $b' = (b_1', \ldots, b_r')$ over B such that $x_i := b_i - b_i' \in mB$ for every i and $< b', n >= 0$. But then $\alpha =< b, n >=< x, n > + < b', n >=< x, n >$ is in mN as required.

3. Tangential flatness under base change

3.1. Strict ideal generators

Let A be a filtered ring, $I \subseteq A$ an ideal, and $x := (x_\lambda)_{\lambda \in \Lambda}$ a family of elements from I. Then x is called a *strict system of generators* of I, if

$$(1) \qquad F_A^d \cap I = \sum_{\lambda \in \Lambda} x_\lambda \cdot F_A^{d-d_\lambda} \qquad \text{for every } d \in \mathbb{N}.$$

Here $d_\lambda := \mathrm{ord}(x_\lambda)$ denotes the order of x_λ with respect to F_A. These identities imply, as one can easily see, that the initial ideal $\mathrm{in}(I)$ is generated by the initial forms $\mathrm{in}(x_\lambda)$,

$$(2) \qquad \mathrm{in}(I) = \mathrm{in}_{F_A}(I) = \sum_{\lambda \in \Lambda} \mathrm{in}(x_\lambda) \cdot G(A).$$

Moreover, a strict system of generators is also a generating system for the ideal I (let $d = 0$ in the definition). A generating system x of I satisfying condition (2) is called a *standard basis* of I.

Every ideal I in a filtered ring has a strict system of generators (hence a standard basis). Too see this, consider the family $x := (x_\lambda)_{\lambda \in \Lambda}$ of all elements of the ideal. Then $x_\lambda \in F_A^d \cap I$ implies that x_λ has order at least d, i.e., the right hand side of (1) contains the term $x_\lambda \cdot F_A^{d-d_\lambda} = x_\lambda \cdot A$ which in turn contains x_λ. Thus, the left hand side of (1) is contained in the right hand side. Since the converse inclusion is trivial, x is a strict system of generators.

It is quite easy to find examples of non-strict generating systems forming a standard basis. However, the next statement shows that the two notions coincide in the situation we are mainly interested in.

3.2. Strictness criterion

Let A be a filtered ring, $I \subseteq A$ an ideal, and $x := (x_\lambda)_{\lambda \in \Lambda}$ a family of elements from I. Assume that the following conditions are satisfied.

1. A is a local ring.
2. F_A is an Artin-Rees filtration.

Then the following conditions are equivalent.

(i) x is a strict system of generators for I.

(ii) x is a standard basis of I.

Proof. Implication (i)⇒(ii) is true without the additional assumptions as we have seen in 3.1. Assume now that condition (ii) is satisfied. Then, the initial form of an arbitrary order d element from I can be written as a linear combination of the $\text{in}(x_\lambda)$'s with homogeneous coefficients. Passing to the representatives in A, we see that,

$$F_A^d \cap I = \sum_{\lambda \in \Lambda} x_\lambda \cdot F_A^{d-d_\lambda} + F_A^{d+1} \cap I.$$

for every d. Iterating these identities, we obtain

$$F_A^d \cap I = \sum_{\lambda \in \Lambda} x_\lambda \cdot F_A^{d-d_\lambda} + F_A^{d+k} \cap I \quad \text{for } k = 0, 1, 2, \cdots.$$

Since F_A is, by assumption, an Artin-Rees filtration, the last term on the right can be replaced by the product $F_A^d \cap I \cdot F_A^{d+k}$, which in turn can be omitted at all by Nakayama's lemma ([Ma86], Corollary to Th.2.2). This proves the claim.

3.3. Liftable syzygies

Let $f : A \to B$ be a ring homomorphism and $x := (x_\lambda)_{\lambda \in \Lambda}$ a family of elements from A. A *syzygy* of x over B is an element

$$b := (b_\lambda)_{\lambda \in \Lambda} \in B^{(\Lambda)} := \bigoplus_{\lambda \in \Lambda} B$$

such that

$$< b, x >:= \sum_{\lambda \in \Lambda} b_\lambda x_\lambda = 0.$$

Note that the syzygies of x in B form a module over B. This module will be denoted by

$$R_B(x) := \{b \in B^{(\Lambda)} | < b, x >= 0\}.$$

Now let $f : A \to B$ be a homomorphism of filtered rings, $x := (x_\lambda)_{\lambda \in \Lambda}$ a family of elements from A, and $u := (u_\lambda)_{\lambda \in \Lambda}$ a syzygy over $G(B)$ of the family $\text{in}(x) := (\text{in}_{F_A}(x_\lambda))_{\lambda \in \Lambda}$ of initial forms. Then u is called a *homogeneous syzygy* of degree d, if

$$u_\lambda \in G(B)(d - d_\lambda) \quad \text{for every } \lambda \in \Lambda, (d_\lambda := \text{ord}_A x_\lambda).$$

We will say that the homogeneous syzygy u of $\text{in}(x)$ is *liftable to a syzygy in B*, if there exists a syzygy $b := (b_\lambda)_{\lambda \in \Lambda}$ of x over B such that

$$u_\lambda = (b_\lambda \bmod F_B^{d-d_\lambda+1}) \quad \text{for every } \lambda \in \Lambda.$$

3.4. Strict homomorphisms of filtered modules

Let A be a ring and $f : (M, F_M) \to (N, F_N)$ a homomorphism of filtered A-modules. Then f is called a *strict homomorphism* if, for every $d \in \mathbb{Z}$,

$$f(F_M^d) = F_N^d \cap \text{Im}(f).$$

(see [Del71, (1.1.6)]). Strict homomorphisms are interesting since, given an exact sequence

$$(M, F_M) \xrightarrow{f} (N, F_N) \xrightarrow{g} (L, F_L)$$

of strict homomorphisms, the associated sequence of graded modules

$$G(M) \xrightarrow{G(f)} G(N) \xrightarrow{G(g)} G(L)$$

is also exact. This can be seen as follows. Equip $\text{Ker}(f)$ and $\text{Im}(f)$ with the induced filtrations $\text{Ker}(f) \cap F_M$ and $\text{Im}(f) \cap F_N$, respectively. Then, since f is strict,

$$\begin{aligned}
\text{Im } G(f) &= \bigoplus_{d \in \mathbb{Z}} f(F_M^d) + F_N^{d+1} / F_N^{d+1} \\
&= \bigoplus_{d \in \mathbb{Z}} F_N^d \cap \text{Im}(f) + F_N^{d+1} / F_N^{d+1} \\
&= G(\text{Im } f)
\end{aligned}$$

and

$$\begin{aligned}
\text{Ker } G(f) &= \bigoplus_{d \in \mathbb{Z}} f^{-1}(F_N^{d+1}) \cap F_M^d / F_M^{d+1} \\
(1) \qquad &= \bigoplus_{d \in \mathbb{Z}} \text{Ker}(f) \cap F_M^d + F_M^{d+1} / F_M^{d+1} \\
&= G(\text{Ker } f)
\end{aligned}$$

As for identity (1) note that $x \in f^{-1}(F_N^{d+1}) \cap F_M^d$ implies $f(x) \in F_N^{d+1} \cap \text{Im}(f) = f(F_M^{d+1})$, hence $x - x' \in \text{Ker}(f) \cap F_M^d$ for some $x' \in F_M^{d+1}$, hence $x \in \text{Ker}(f) \cap F_M^d + F_M^{d+1}$.

Since the above identities are equally valid with f replaced by g and since, by assumption, $\text{Im } f = \text{Ker } g$,

$$\text{Im } G(f) = G(\text{Im } f) = G(\text{Ker } g) = \text{Ker } G(g).$$

3.5. Example of a strict morphism

Let A be a filtered ring, $I \subseteq A$ an ideal, and $x := (x_\lambda)_{\lambda \in \Lambda}$ a family of elements generating I. Equip the direct sum $M := A^{(\Lambda)} := \bigoplus_{\lambda \in \Lambda} A$ with the filtration

$$F_M^d := \bigoplus_{\lambda \in \Lambda} F_A^{d - d_\lambda}, \quad d_\lambda := \text{ord } x_\lambda,$$

and consider the homomorphism of filtered A-modules,

$$f = <?, x> : M \to A, \quad a = (a_\lambda)_{\lambda \in \Lambda} \mapsto <a, x> := \sum_{\lambda \in \Lambda} a_\lambda x_\lambda$$

Then

$$f(F_M^d) = \sum_{\lambda \in \Lambda} x_\lambda \cdot F_A^{d-d_\lambda} \quad \text{and} \quad \text{Im}(f) = I.$$

In particular, $f = <?, x>$ is a strict homomorphism if and only if the elements x_λ form a strict system of generators for I.

3.6. The exact sequence associated with a generating system

Let $f : A \to B$ be a homomorphism of filtered rings, $I \subseteq A$ an ideal, and $x = (x_\lambda)_{\lambda \in \Lambda}$ a generating system of I. Then there is an exact sequence

$$0 \to R_B(x) \xrightarrow{\alpha} B^{(\Lambda)} \xrightarrow{<?,x>} B \xrightarrow{\beta} B/IB \to 0$$

which is called the *exact sequence associated with x over B*. Here $B^{(\Lambda)}$ is the free B-module

$$B^{(\Lambda)} := \bigoplus_{\lambda \in \Lambda} B$$

and the map in the middle is the scalar product with x,

$$<b, x> := \sum_{\lambda \in \Lambda} b_\lambda x_\lambda$$

The injection on the left and the surjection on the right are the canonical embedding and the canonical homomorphism, respectively. All the modules in the above exact sequence are naturally equipped with filtrations, such that the homomorphisms in the sequence are homomorphisms of filtered modules,

$$F_{B/IB} := F_B \cdot B/IB$$
$$F_{B^{(\Lambda)}}^d := \bigoplus_{\lambda \in \Lambda} F_B^{d-d_\lambda}, \quad d_\lambda := \text{ord } x_\lambda,$$
$$F_{R_B(x)}^d := F_{B^{(\Lambda)}}^d \cap R_B(x)$$

Moreover, the two outer homomorphisms α and β of the sequence are strict with respect to these filtrations as one can see directly from the definition of strictness in 3.4. The homomorphism $<?, \text{in}(x)>$ in the middle is strict in case that x is a strict system of generators for I (see 3.5) and $(B, F_B) := (A, F_A)$. So for strict generating systems x of I, the induced sequence of graded homomorphisms

$$0 \to G(R_A(x)) \to G(A^{(\Lambda)}) \xrightarrow{<?,\text{in}(x)>} G(A) \to G(A/I) \to 0$$

is also exact. This will turn out to be important later.

3.7. Exactness of the associated graded sequence

Let $f : A \rightarrow B$ be a homomorphism of filtered rings, $I \subseteq A$ an ideal, and

$$x = (x_\lambda)_{\lambda \in \Lambda}$$

a generating system of I. Consider the complex

(1) $\qquad 0 \rightarrow G(R_B(x)) \rightarrow G(B^{(\Lambda)}) \xrightarrow{\ <?,\mathrm{in}(x)>\ } G(B) \rightarrow G(B/IB) \rightarrow 0$

obtained applying the functor $G(?)$ to the exact sequence associated with x over B (cf 3.6). The following assertions are equivalent.

(i) The complex is an exact sequence.

(ii) The complex is exact at $G(B^{(\Lambda)})$.

(iii) Every homogeneous syzygy over $G(B)$ of $\mathrm{in}(x) := (\mathrm{in}_{F_A}(x_\lambda))_{\lambda \in \Lambda}$ is liftable to a syzygy of x over B.

Proof. (iii)$\Rightarrow$(i). We have to prove the sequence is exact at $G(B^{(\Lambda)})$ and $G(B)$, for, it is always exact at the other locations by 3.6 and 3.4. Let

$$u = (u_\lambda)_{\lambda \in \Lambda} \in G(B^{(\Lambda)}) = \bigoplus_{\lambda \in \Lambda} G(B)[-d_\lambda], \quad d_\lambda := \mathrm{ord}(x_\lambda),$$

be in the kernel of $<?, \mathrm{in}(x) >$. We want to prove that u is in $G(R_B(x))$. For this we may assume that u is homogeneous of degree, say, d. By assumption (iii) there exists some syzygy $b = (b_\lambda)_{\lambda \in \Lambda} \in R_B(x)$ such that, for every $\lambda \in \Lambda$,

$$u_\lambda = (b_\lambda \bmod F_B^{d+1-d_\lambda}).$$

But then $u = (b \bmod F_{B^{(\Lambda)}}^{d+1})$ can be considered as a homogeneous element of degree d in $G(R_B(x))$. For, since (1) is exact at $G(R_B(x))$, a homogeneous element of $G(B^{(\Lambda)})$ is in $G(R_B(x))$ if and only if it can be represented by an element of $R_B(x)$. This shows, the sequence (1) is exact at $G(B^{(\Lambda)})$. We have yet to prove that the kernel $\mathrm{in}(IB)$ of the canonical surjection $G(B) \rightarrow G(B/IB)$ equals the image of the mapping

$$G(B^{(\Lambda)}) = \bigoplus_{\lambda \in \Lambda} G(B)[-d_\lambda] \xrightarrow{\ <?,\mathrm{in}(x)>\ } G(B), \quad u \mapsto < u, \mathrm{in}(x) >,$$

i.e.,

(2) $\qquad\qquad\qquad \mathrm{in}(y) \in \sum_{\lambda \in \Lambda} \mathrm{in}(x_\lambda) \cdot G(B)$

for every $y \in IB$ of finite order. Write

$$y = \sum_{\lambda \in \Lambda} x_\lambda b_\lambda, \quad b_\lambda \in B,$$

and define

$$d := \min\{d_\lambda + \operatorname{ord}(b_\lambda) | \lambda \in \Lambda\}.$$

We are going to prove (2) by descending induction on d. In case d is large, say

$$\operatorname{ord}(y) \leq d \quad (\leq d_\lambda + \operatorname{ord}(b_\lambda) \text{ for every } \lambda)$$

the definition of d implies that one has even equality, $\operatorname{ord}(y) = d$. In particular, it is possible to write

$$\operatorname{in}(y) = \sum_{\lambda \in \Lambda} \operatorname{in}(x_\lambda) \cdot (b_\lambda \bmod F_B^{d-d_\lambda+1}) \in \sum_{\lambda \in \Lambda} \operatorname{in}(x_\lambda) \cdot G(B)$$

i.e., (2) is true. Now assume $\operatorname{ord}(y) > d$. Then

$$0 = (y \bmod F_B^{d+1}) = \sum_{\lambda \in \Lambda} \operatorname{in}(x_\lambda) \cdot (b_\lambda \bmod F_B^{d-d_\lambda+1}),$$

i.e., the elements $u_\lambda := (b_\lambda \bmod F_B^{d-d_\lambda+1})$ form a homogeneous syzygy u of $\operatorname{in}(x)$ over $G(B)$. By assumption, u can be lifted to a syzygy of x, i.e., there is a family

$$b' := (b'_\lambda)_{\lambda \in \Lambda}$$

of elements from B such that $< x, b' >= 0$ and $b_\lambda - b'_\lambda \in F_B^{d+1-d_\lambda}$ for every $\lambda \in \Lambda$. Therefore,

$$y = < x, b > = < x, b - b' > = \sum_{\lambda \in \Lambda} x_\lambda \cdot (b_\lambda - b'_\lambda)$$

and

$$d < \min\{d_\lambda + \operatorname{ord}(b_\lambda - b'_\lambda) | \lambda \in \Lambda\}.$$

Now (2) follows from the induction hypothesis.

(i)$\Rightarrow$(ii). Trivial.

(ii)$\Rightarrow$(iii). Let $u = (u_\lambda)_{\lambda \in \Lambda}$ be a homogeneous syzygy of degree d of $\operatorname{in}(x)$ over $G(B)$. Then u is in the kernel of $<?, \operatorname{in}(x) >$. Since, by assumption, the sequence (1) is exact at $G(B^{(\Lambda)})$, there exists an element

$$b = (b_\lambda)_{\lambda \in \Lambda} \in R_B(x)$$

such that $u_\lambda = (b_\lambda \bmod F_B^{d-d_\lambda+1})$ for every $\lambda \in \Lambda$. Hence u can be lifted to a syzygy b of x.

3.8. Tangential flatness under surjective base change

Let $f : A \to B$ be a homomorphism of filtered rings, and $I \subseteq A$ an ideal. Suppose that f is tangentially flat. Then

(i) $f \otimes_A A/I : A/I \to B/IB$ is tangentially flat.

(ii) $\mathrm{in}(IB) = \mathrm{in}(I)\,G(B)$.

Proof. Choose some strict generating system $x = (x_\lambda)_{\lambda \in \Lambda}$ of I, define as usual $d_\lambda := \mathrm{ord}_A x_\lambda$, and consider the complex

$$0 \to G(R_A(x)) \to G(A^{(\Lambda)}) \xrightarrow{\;<?,\mathrm{in}(x)>\;} G(A) \to G(A/I) \to 0$$

obtained applying the functor $G(?)$ to the exact sequence associated with x over A. Since x is strict, this complex is even an exact sequence by 3.6. Apply the functor $\otimes_{G(A)} G(B)$ and insert the resulting exact sequence into a commutative diagram as follows.

$$
\begin{array}{ccc}
0 & & 0 \\
\downarrow & & \downarrow \\
G(R_A(x)) \otimes G(B) & \xrightarrow{\;\delta\;} & G(R_B(x)) \\
\downarrow & & \downarrow \\
G(A^{(\Lambda)}) \otimes G(B) & \xrightarrow{\;\gamma\;} & Gr(B^{(\Lambda)}) \\
\downarrow {\scriptstyle <?,\mathrm{in}(x)>} & & \downarrow {\scriptstyle <?,\mathrm{in}(x)>} \\
G(B) & \stackrel{\beta}{=\!=} & G(B) \\
\downarrow & & \downarrow \\
G(B)/\mathrm{in}(I)\,G(B) & \xrightarrow{\;\alpha\;} & G(B/IB) \\
\downarrow & & \downarrow \\
0 & & 0
\end{array}
$$

Here the right hand side column is obtained applying the functor $G(?)$ to the exact sequence associated with x over B (cf 3.6). The homomorphism α is the canonical surjection, and γ the canonical isomorphism,

$$G(A^{(\Lambda)}) \otimes G(B) \cong \left(\bigoplus_{\lambda \in \Lambda} G(A)[-d_\lambda] \right) \otimes G(B) \longrightarrow \bigoplus_{\lambda \in \Lambda} G(B)[-d_\lambda] \cong G(B^{(\Lambda)})$$

mapping $\mathrm{in}((a_\lambda)_{\lambda \in \Lambda}) \otimes \mathrm{in}(b)$ to $\mathrm{in}((a_\lambda b)_{\lambda \in \Lambda})$. The existence of δ can be seen as follows. Let $\mathrm{in}(a) \in G(R_A(x))$ and $\mathrm{in}(b) \in G(B)$ be homogeneous elements. In particular, let $a = (a_\lambda)_{\lambda \in \Lambda} \in R_A(x)$ and $b \in B$. It will be sufficient to show that the image of $\mathrm{in}(a) \otimes \mathrm{in}(b) \in G(A^{(\Lambda)}) \otimes G(B)$ under γ is even in $G(R_B(x))$. Since $a \in R_A(x)$, $< a, x >= 0$, hence

$$\gamma(\mathrm{in}(a) \otimes \mathrm{in}(b)) = \gamma(\mathrm{in}((a_\lambda)_{\lambda \in \Lambda}) \otimes \mathrm{in}(b)) = \mathrm{in}((a_\lambda b)_{\lambda \in \Lambda})$$

satisfies

$$< (a_\lambda b)_{\lambda \in \Lambda}, x >=< a, x > \cdot b = 0.$$

Thus $\gamma(\text{in}(a) \otimes \text{in}(b))$ can be represented by an element $(a_\lambda b)_{\lambda \in \Lambda}$ of $R_B(x)$, as required, and we have proved the existence of the homomorphism δ above. Since γ and β are isomorphisms, the right hand side column of the diagram must be exact at $G(B^{(\Lambda)})$. By 3.7 the column is exact everywhere, hence α is an isomorphism by the five lemma. Therefore $\text{in}(IB) = \text{in}(I)\,G(B)$, i.e., assertion (ii) is true. Further, since $G(B)$ is $G(A)$-flat, the ring

$$G(B/IB) \cong G(B)/\text{in}(IB) = G(B)/\text{in}(I)\,G(B)$$

is flat over $G(A/I) = G(A)/\text{in}(I)$, which is just the claim of (i).

3.9. Tangential flatness under polynomial extension

Let $f : A \to B$ be a homomorphism of filtered rings, $X := (X_\lambda)_{\lambda \in \Lambda}$ a family of indeterminates, and $(d_\lambda)_{\lambda \in \Lambda}$ a family with $d_\lambda \in \mathbb{N} \cup \{\infty\}$. Define

$A' := A[X],$
$B' := B[X],$
$F_{A'} :=$ the filtration generated over $F_A A'$ by X such that $\text{ord}\, X_\lambda \geq d_\lambda$
$F_{B'} :=$ the filtration generated over $F_B B'$ by X such that $\text{ord}\, X_\lambda \geq d_\lambda$

Then, if f is tangentially flat, the same is true for the homomorphism

$$f' : (A', F_{A'}) \to (B', F_{B'})$$

induced by f.

In case the family of indeterminates X is finite, the statement continues to be true when the polynomial rings A' and B' above are replaced by the corresponding power series rings,

$$A' := A[[X]], \quad B' := B[[X]].$$

See 1.11 for the definition of a filtration generated by a family.

Proof. By definition,

(1) $$F_{A'}^d = \sum_{(i,\lambda) \in \mathbb{N}^n \times \Lambda^n} F_A^{d - <i, d_\lambda>} \cdot X_\lambda^i$$

where n runs over the non-negative integers and

$X_\lambda^i := (X_{\lambda_1})^{i_1} \cdot \ldots \cdot (X_{\lambda_n})^{i_n}$
$d_\lambda := (d_{\lambda_1}, \ldots, d_{\lambda_n})$
$< i, d_\lambda > := i_1 \cdot d_{\lambda_1} + \ldots + i_n \cdot d_{\lambda_n}$

if $i = (i_1, \ldots, i_n)$ and $\lambda = (\lambda_1, \ldots, \lambda_n)$. Since the right hand side of (1) is a direct sum, we get

$$G(A')(d) = \bigoplus_{(i,\lambda) \in \mathbb{N}^n \times \Lambda^n} (F_A^{d - <i,d_\lambda>}/F_A^{d+1 - <i,d_\lambda>}) \cdot X_\lambda^i$$

$$G(A') = \bigoplus_{(i,\lambda) \in \mathbb{N}^n \times \Lambda^n} G(A)[- < i, d_\lambda >] \cdot X_\lambda^i$$

Let X' be the set of indeterminates X_λ such that d_λ is finite. Then $G(A')$ is just the polynomial ring

$$G(A') = G(A)[X'],$$

where $X_\lambda \in X'(\lambda \in \Lambda)$ is a variable of degree d_λ in $G(A')$. Similarly we obtain,

$$G(B') = G(B)[X'].$$

Since $G(B)$ is flat over $G(A)$, the ring $G(B') \cong G(B) \otimes_{G(A)} G(A')$ is flat over $G(A')$.

In the power series case the direct sum decomposition (1) becomes a direct product, hence in the above formula for $G(A')(d)$, the direct sum on the right should be replaced by the corresponding direct product, at least at first glance. However, due to the finiteness assumption on X, the number of non-zero direct factors in the decomposition is finite, so that the direct product is also a direct sum. The same calculations as above give now the claim in the case of power series.

3.10. Tangential flatness and increased base filtration

Let $f : A \to B$ be a tangentially flat homomorphism of filtered rings, and F a ring filtration of A containing F_A. Then the homomorphism

$$f : (A, F) \to (B, F_B + f_* F)$$

is also tangentially flat.

Proof. Let $x := (x_\lambda)_{\lambda \in \Lambda}$ be a family generating F over F_A such that

$$\mathrm{ord}\, x_\lambda \geq d_\lambda (\lambda \in \Lambda).$$

Such a family exists by 1.11. Define

$$A' := A[X], \quad B' := B[X]$$

where $X := (X_\lambda)_{\lambda \in \Lambda}$ is a family of indeterminates, and

$F_{A'} :=$ the filtration generated over $F_A \cdot A'$ by X such that $\mathrm{ord}\, X_\lambda \geq d_\lambda$

$F_{B'} :=$ the filtration generated over $F_B \cdot B'$ by X such that $\mathrm{ord}\, X_\lambda \geq d_\lambda$

Then, as a filtered ring, (A, F) is just the factor ring of $(A', F_{A'})$ modulo the ideal I of A' generated by the linear polynomials $X_\lambda - x_\lambda$. Similarly, $(B, F_B + f_* F)$ is the factor of $(B', F_{B'})$ modulo IB'. Note that,

$$(F_B + f_* F)^d = \sum_{i+j=d} F^i F_B^j \quad \text{(by 1.6 and 1.10)}$$

$$= \sum_{i+j=d, (k,\lambda) \in \mathbb{N}^n \times \Lambda^n} F_A^{i-<k,d_\lambda>} \cdot x_\lambda^k \cdot F_B^j \quad \text{(by 1.11)}$$

$$= \sum_{(k,\lambda) \in \mathbb{N}^n \times \Lambda^n} x_\lambda^k \cdot F_B^{d-<k,d_\lambda>},$$

i.e., the filtration $F_B + f_* F$ is generated over F_B by the family $(x_\lambda)_{\lambda \in \Lambda}$ such that $\operatorname{ord} x_\lambda \geq d_\lambda$.

To prove that f is tangentially flat, it is (by 3.8) sufficient to show that $(B', F_{B'})$ is tangentially flat over $(A', F_{A'})$. But this was proved in 3.9.

3.11. Convention

From now on we will almost exclusively restrict to local rings and local homomorphisms, though many of the results below have generalizations to the non-local situation. The local context is the one we are interested in, and the restriction to this case will considerably simplify most statements and their proofs.

Unless stated otherwise we will assume that *all ring filtrations are separated and cofinite Artin-Rees filtrations*. In particular, the topology defined by the filtration F_A of a local ring (A, m) will be the m-adic topology (see 1.15(vi)). A filtration as originally defined in 1.2 (i.e., one which is not necessarily separated or cofinite or an Artin-Rees filtration) will be called a *prefiltration*.

Factor rings A/I of filtered rings A will be automatically equipped with the induced filtration $F_A \cdot A/I$ (see 1.10).

Remark

Note that the above conventions imply that for every filtered local ring (A, m) and every proper ideal $I \subseteq A$ the following conditions are satisfied.

 (i) Each homogeneous part $G(A)(d)$ of $G(A)$ has finite length over A, hence is annihilated by some power of the degree zero part $\operatorname{in}(I)(0)$ of the initial ideal $\operatorname{in}(I)$. In particular, condition 2 of the graded Nakayama lemma 2.1 is satisfied for the submodules of $G(A)$ and the ideal $\operatorname{in}(I)$.

 (ii) The $\operatorname{in}(I)$-adic topology of $G(A)$ is separated, i.e.,

$$\bigcap_{n=0}^{\infty} \operatorname{in}(I)^n = 0$$

(iii) The ideal $\operatorname{in}(I)$ has even the stronger property that, for every $d \in \mathbb{N}$ there is some $n \in \mathbb{N}$ such that $\operatorname{in}(I)^n$ doesn't have any homogeneous elements of degree less than d,
$$\operatorname{in}(I)^n(< d) = 0$$

Note that since $\operatorname{in}(I) \subseteq \operatorname{in}(I)(0) + G^+(A)$,

$$\operatorname{in}(I)^n \subseteq (\operatorname{in}(I)(0) + G^+(A))^n$$
$$= \sum_{i=0}^{n} (\operatorname{in}(I)(0))^{n-i} G^+(A)^i)$$
$$\subseteq \bigoplus_{k \in \mathbb{N}} \operatorname{in}(I)(0)^{n-k} G(k) = 0.$$

For n large enough, the first d homogeneous parts of the last module are zero by assertion (i).

3.12. Minimality of tangentially flat filtrations

Let $f : (A, m) \to (B, n)$ be a homomorphism of filtered local rings and let F be an A-module filtration of B compatible with F_A and such that

1. $F \subseteq F_B$.
2. $\rho_* F = \rho_* F_B$ where $\rho : B \to B/mB$ is the natural homomorphism and ρ_* denotes the direct image in the sense of 1.12 (rather than 1.10).
3. Each F^d is closed in the topology defined by F_B.

Then, if f is tangentially flat, $F = F_B$.

In particular, the ring filtrations of B making f tangentially flat are minimal among all filtration of B inducing a given filtration on the special fibre.

Proof. Let B' be the A-module B equipped with the filtration F. Then there is an exact sequence of modules over $G(A)$,

$$G(B') \xrightarrow{\phi} G(B) \to \operatorname{Coker}(\phi) \to 0$$

where ϕ is induced by the inclusions $F^d \subseteq F_B^d$. Since $G(B)$ is flat over $G(A)$, there is a natural isomorphism $G(B)/\operatorname{in}(m)G(B) \cong G(B/mB)$ by 3.8, hence a commutative diagram of natural homomorphisms

$$
\begin{array}{ccc}
G(B') & \to & G(B')/\operatorname{in}(m)\,G(B') \\
& & \quad\downarrow \phi \otimes_{G(A)} G(A)/\operatorname{in}(m) \\
& & G(B)/\operatorname{in}(m)\,G(B) \\
\downarrow & & \downarrow \cong \\
G(B'/mB') & \cong & G(B/mB)
\end{array}
$$

which shows that $\phi \otimes_{G(A)} G(A)/\operatorname{in}(m)$ is surjective, i.e.,

$$\operatorname{Coker}(\phi) = \operatorname{in}(m) \cdot \operatorname{Coker}(\phi).$$

The filtration F_B is cofinite by convention (see 3.11), hence the homogeneous parts of $G(B)$ are annihilated by sufficiently high powers of m, and the same is true for the homogeneous parts of its $G(A)$-linear image $\operatorname{Coker}(\phi)$. Since $\operatorname{in}(m)(0)$ is a factor of m, the graded Nakayama lemma 2.1 implies $\operatorname{Coker}(\phi) = 0$, i.e., ϕ is surjective. Therefore,

$$F_B^d = F^d + F_B^{d+1} \quad \text{for every } d.$$

Iterating these identities, we get $F_B^d = F^d + F_B^k$ for every $k \geq d$, hence by condition 3,

$$F_B^d = \bigcap_{k \in \mathbb{N}} (F^d + F_B^k) = F^d$$

for every d, i.e., $F = F_B$ as required.

3.13. Initial forms of extended ideals

Let $f : (A, m) \to (B, n)$ be a homomorphism of filtered local rings and $I \subseteq A$ an ideal. Then the following assertions are equivalent.

(i) $\operatorname{in}(IB) = \operatorname{in}(I) G(B)$.
(ii) $IB \cap F_B^d = \sum_{i+j=d} (I \cap F_A^i) \cdot F_B^j$ for every $d \in \mathbb{N}$.

Proof. Let

$$F(d) := \sum_{i+j=d} (I \cap F_A^i) \cdot F_B^j$$

be the expression on the right hand side of (ii). Then condition (i) is equivalent to the inclusions

$$IB \cap F_B^d \subseteq F(d) + F_B^{d+1},$$

hence to

$$IB \cap F_B^d = F(d) + IB \cap F_B^k \text{ for } k \geq d.$$

Since F_B is an Artin-Rees filtration, $IB \cap F_B^k \subseteq I \cdot F_B^d \subseteq F(d)$ for large $k = k(d)$, i.e., the second term on the right can be omitted. So (i) and (ii) are equivalent.

Remark

We introduce now a short exact sequence that will play a crucial role in this paper. A careful analysis of this sequence will uncover the relation of tangential flatness to the behavior of the local Hilbert series under local extensions.

3.14. The basic exact sequence

Let $f : (A, m) \to (B, n)$ be a homomorphism of filtered local rings. Then there is, for arbitrary $d, i \in \mathbb{N}$ an exact sequence

$$0 \to D(d, i) \to \mathrm{G}_{F_A}(B)(i) \otimes_B B(d - i) \to \mathrm{G}_{F_A}(B(d))(i) \to 0$$

with

$$D(d, i) := (F_A^i B \cap F_B^{d+1})/(F_A^i \cdot F_B^{d-i+1} + F_A^{i+1} B \cap F_B^{d+1})$$

Moreover, the following conditions are equivalent.

(i) $\mathrm{G}_{F_A}(B)(i) \otimes_B B(d - i) \to \mathrm{G}_{F_A}(B(d))(i)$ is an isomorphism for arbitrary d and i with $i \leq d$.

(ii) $\mathrm{in}(F_A^i B) = \mathrm{in}(F_A^i) \, \mathrm{G}(B)$ for every i.

See 1.13 for terminology.

Proof. By definition $B(d) = B/F_B^{d+1}$ hence

$$\mathrm{G}_{F_A}(B(d))(i) \cong (F_A^i B + F_B^{d+1})/(F_A^{i+1} B + F_B^{d+1})$$
$$\cong F_A^i B/(F_A^{i+1} B + F_A^i B \cap F_B^{d+1}).$$

Comparing this with

$$\mathrm{G}_{F_A}(B)(i) \otimes_B B(d - i) = (F_A^i B/F_A^{i+1} B) \otimes_B B(d - i)$$
$$\cong F_A^i B/(F_A^{i+1} B + F_A^i F_B^{d-i+1}),$$

we see that there is a surjection $\mathrm{G}_{F_A}(B)(i) \otimes_B B(d - i) \to \mathrm{G}_{F_A}(B(d))(i)$ with kernel equal to

$$(F_A^{i+1} B + F_A^i B \cap F_B^{d+1})/(F_A^{i+1} B + F_A^i F_B^{d-i+1})$$
$$\cong F_A^i B \cap F_B^{d+1}/(F_A^{i+1} B \cap F_B^{d+1} + F_A^i F_B^{d-i+1})$$
$$= D(d, i).$$

This gives the exact sequence we are looking for. In order to prove the second part of the claim, note that by 3.13, condition (ii) is equivalent to the inclusions,

$$F_A^i B \cap F_B^d \subseteq \sum_{r+s=d, r \geq i} F_A^r F_B^s,$$

which in turn are equivalent to

$$F_A^i B \cap F_B^d \subseteq F_A^i F_B^{d-i} + F_A^{i+1} B \cap F_B^d.$$

The latter inclusions are equivalent to the identities $D(d-1, i) = 0$. We have proved, condition (i) is satisfied if and only if so is condition (ii).

Remark

The theorem below can be considered as a converse of 3.8. It states that $f : A \to B$ is tangentially flat if and only if so is $f \otimes A/I$ with $I := F_A^1$ and certain natural surjections are bijective (which are very close to the surjections of the basic exact sequence above). We will prove later that these bijectivity conditions can be replaced by the condition $\mathrm{in}(IB) = \mathrm{in}(I)\,\mathrm{G}(B)$. As an immediate consequence of 3.15 we will see that tangential flatness is an 'infinitesimal' property (see 3.16).

3.15. Tangential flatness and bijectivity of certain surjections

Let $f : (A, m) \to (B, n)$ be a homomorphism of filtered local rings. Then the following conditions are equivalent.

(i) f is tangentially flat.

(ii) $f \otimes_A A(0) : A/F_A^1 \to B/F_A^1 B$ is tangentially flat and the surjections

$$G(A)(i) \otimes_A B(d-i) \to \mathrm{G}_{F_A}(B(d))(i)$$

induced by multiplication are bijective for arbitrary d and i with $d \geq i$.

Proof. First note that multiplication induces A-bilinear maps,

$$G(A)(i) \times B(d-i) \to \mathrm{G}_{F_A}(B(d))(i) = (F_A^i B + F_B^{d+1})/(F_A^{i+1} B + F_B^{d+1}),$$

i.e., the homomorphisms of (ii) are well-defined. They are obviously surjective.

(i)$\Rightarrow$(ii). Tangential flatness of f implies by 3.8 that $f \otimes A/F_A^1$ is tangentially flat. So we have just to prove bijectivity of the homomorphisms in (ii). We will use the notation

$$B_0 := B/F_A^1 B, \quad A_0 := A/F_A^1.$$

Consider the natural isomorphism

$$G(A) \to G^* := \bigoplus_{i \in \mathbb{N}} \mathrm{in}(F_A^i)/\mathrm{in}(F_A^{i+1})$$

which maps the homogeneous element $x \in G(A)(d)$ into $x + \mathrm{in}(F_A^{d+1})$ and defines a new $G(A)$-module structure on $G(A)$ such that $G(A)$ is annihilated by $\mathrm{G}^+(A) = \mathrm{in}(F_A^1)$. With this module structure,

$$
\begin{aligned}
G(A) \otimes_{A_0} G(B_0) \; &\cong \; G^* \otimes_{A_0} G(B)/\mathrm{G}^+(A)\,\mathrm{G}(B) \\
&\cong \; G^* \otimes_{G(A)/\mathrm{G}^+(A)} G(B)/\mathrm{G}^+(A)\,\mathrm{G}(B) \\
&\cong \; G^* \otimes_{G(A)} G(B) \\
&\cong \; \bigoplus_{i \in \mathbb{N}} \mathrm{in}(F_A^i)\,\mathrm{G}(B)/\mathrm{in}(F_A^{i+1})\,\mathrm{G}(B) \\
&\cong \; \bigoplus_{i \in \mathbb{N}} \mathrm{in}(F_A^i B)/\mathrm{in}(F_A^{i+1} B).
\end{aligned}
$$

Note that the first and the last two isomorphisms come from the fact that $G(B)$ is flat over $G(A)$. Restrict this isomorphism to $G(A)(i) \otimes G(B_0)(d - i)$ to get a bijection

$$G(A)(i) \otimes G(B_0)(d - i) \to (\mathrm{in}(F_A^i B)/\mathrm{in}(F_A^{i+1}B))(d)$$

and include the latter into a commutative diagram.

$$
\begin{array}{ccc}
0 & & 0 \\
\downarrow & & \downarrow \\
G(A)(i) \otimes G(B_0)(d - i) & \xrightarrow{\cong} & (\mathrm{in}(F_A^i B)/\mathrm{in}(F_A^{i+1}B))(d) \\
\downarrow & & \downarrow \\
G(A)(i) \otimes B(d - i) & \longrightarrow & G_{F_A}(B(d))(i) \\
\downarrow & & \downarrow \\
G(A)(i) \otimes B(d - i - 1) & \longrightarrow & G_{F_A}(B(d - 1))(i) \\
\downarrow & & \downarrow \\
0 & & 0
\end{array}
$$

The diagram is constructed as follows. The left hand side column is obtained applying the functor $G(A)(i) \otimes_A$ to the short exact sequence

$$0 \to G(B_0)(j) \to B_0(j) \to B_0(j - 1) \to 0.$$

with $j = d - i$. Note that $G(A)(i) \otimes B(j) \cong G(A)(i) \otimes B_0(j)$, since $G(A)(i)$ is annihilated by F_A^1. Moreover, the left hand side column is exact. To see this, it will be sufficient to show that $B_0(j-1)$ is flat over A_0. In view of the above short exact sequence, it is enough to prove that each $G(B_0)(j)$ is A_0-flat. But this is obvious, since $G(B_0)$ is A_0-flat. In order to get a description of the column on the right, calculate the module in the upper right corner of the diagram. Since,

$$\mathrm{in}(F_A^i B)(d) = F_A^i B \cap F_B^d \bmod F_B^{d+1} = F_A^i B(d) \cap G(B)(d)$$

one can identify

$$
\begin{aligned}
(\mathrm{in}(F_A^i B)/\mathrm{in}(F_A^{i+1}B))(d) &= F_A^i B(d) \cap G(B)(d) + F_A^{i+1}B(d)/F_A^{i+1}B(d) \\
&= \mathrm{in}_{F_A}(G(B)(d))(i) \quad (\subseteq G_{F_A}(B(d))(i)),
\end{aligned}
$$

with the degree i part of the initial ideal $\mathrm{in}_{F_A}(I)$ of $I := G(B)(d)(\subseteq B(d))$ in $G_{F_A}(B(d))$ (cf 1.12(2)). Let the right hand side column of the diagram be, up to these identifications, the degree i part of the defining exact sequence of $\mathrm{in}_{F_A}(I)$ (see 1.12(3)). The two lower horizontal mappings of the diagram are just two of the mappings of condition (ii). We want to prove that these mappings are bijective. For $d = 0$ the bottom horizontal map is trivially an isomorphism, since $B(d - i - 1) = B(d - 1) = 0$. Moreover, by the five lemma, the horizontal map in the middle is bijective, if so is the one at the bottom. Thus the claim follows by induction on d.

(ii)$\Rightarrow$(i). As above let

$$A_0 := A/F_A^1 \text{ and } B_0 := B/F_A^1 B.$$

By assumption $G(B_0)$ is flat over $G(A_0) = A_0$, hence $G(B_0)(j)$ is A_0-flat for every j. From the exact sequences

$$0 \to G(B_0)(j) \to B_0(j) \to B_0(j-1) \to 0$$

we deduce that $B_0(j)$ is A_0-flat for every $j \in \mathbb{Z}$ (which is obvious if j is negative since $B_0(j) = 0$ for $j < 0$). Consider the commutative diagram with exact columns,

$$
\begin{array}{ccc}
0 & & 0 \\
\downarrow & & \downarrow \\
G(A)(i) \otimes G(B_0)(d-i) & \to & \mathrm{in}(G(B)(d))(i) \\
\downarrow & & \downarrow \\
G(A)(i) \otimes B_0(d-i) & \to & G_{F_A}(B(d))(i) \\
\downarrow & & \downarrow \\
G(A)(i) \otimes B_0(d-i-1) & \to & G_{F_A}(B(d-1))(i) \\
\downarrow & & \downarrow \\
0 & & 0
\end{array}
$$

Here the left hand side column is obtained via the functor $G(A)(i) \otimes_{A_0}$ from the above short exact sequence of flat modules. The column on the right is the degree i part of the exact sequence defining the ideal of initial forms $\mathrm{in}(G(B)(d))$(see 1.12(3)). The horizontal map in the middle of the diagram is the composition

$$G(A)(i) \otimes B_0(d-i) \cong G(A)(i) \otimes B(d-i) \to G_{F_A}(B(d))(i)$$

where the second map is the one in the formulation of the theorem and the isomorphism on the left comes from the fact that F_A^1 annihilates $G(A)(i)$. The lower horizontal is obtained analogously. By assumption (ii), the horizontal maps in the middle and at the bottom are isomorphisms, hence so are the maps at the top of the diagram. These latter maps with $i + j = d$ can be put together to give a bijection

$$(G(A) \otimes G(B_0))(d) \to \mathrm{in}(G(B)(d)).$$

Here

$$\mathrm{in}(G(B)(d)) = \bigoplus_{i \in \mathbb{N}} F^i(d)/F^{i+1}(d), \quad F^i(d) := G(B)(d) \cap F_A^i B(d).$$

Define

$$F^i := \bigoplus_{d \in \mathbb{N}} F^i(d) \quad (\subseteq G(B)).$$

Then

$$G_F(G(B)) = \bigoplus_{d \in \mathbb{N}} \bigoplus_{i \in \mathbb{N}} F^i(d)/F^{i+1}(d) = \bigoplus_{d \in \mathbb{N}} \mathrm{in}(G(B)(d)) \cong G(A) \otimes_{A_0} G(B_0).$$

Since $G(B_0)$ is A_0-flat by assumption(ii), we deduce that $G_F(G(B))$ is flat over $G(A)$, hence $G_F(G(B))(i)$ is $G(A)$-flat for every i. From the exact sequences

$$0 \to G_F(G(B))(i) \to G(B)/F^{i+1} \to G(B)/F^i \to 0$$

45

we see inductively that $G(B)/F^i$ is $G(A)$-flat (note that this is the zero module for $i = 0$). Since

$$F^i \subseteq G(B)(\geq i) = \bigoplus_{d=i}^{\infty} G(B)(d),$$

this means that $G(B)$ is $G(A)$-flat by 2.3.

3.16. Tangential flatness as an infinitesimal Property

Let $f : (A, m) \to (B, n)$ be a homomorphism of filtered local rings. Then the following conditions are equivalent.

(i) f is tangentially flat.

(ii) $f \otimes_A A(d) : A/F_A^{d+1} \to B/F_A^{d+1}B$ is tangentially flat for every d.

Proof. Implication (i)$\Rightarrow$(ii) follows from 3.8. So assume $f \otimes_A A(d)$ is tangentially flat for every d. By 3.15 it will be sufficient to show that the natural surjection

$$(1) \qquad\qquad G(A)(i) \otimes_A B(d - i) \to G_{F_A}(B(d))(i)$$

is bijective for arbitrary d and i with $d \geq i$. Note that for fixed d and i this map doesn't change when f is replaced by $f \otimes_A A(d)$. Since the homomorphism $f \otimes_A A(d)$ is tangentially flat by assumption, (1) is bijective by 3.15.

3.17. Tangential flatness under residually rational base change

Let $f : (A, m) \to (B, n)$ and $g : (A, m) \to (A', m')$ be homomorphisms of filtered local rings, where g induces an isomorphism on the residue class fields. Then, if f is tangentially flat, the same is true for

$$f' := f \otimes_A A' : A' \to B' := B \otimes_A A'.$$

provided B' is equipped with the filtration

$$F_{B'} := f'_* F_{A'} + h_* F_B$$

where $h : B \to B'$ denotes the natural map $b \mapsto b \otimes 1$.

Proof. By 3.16 it is sufficient to show that $f' \otimes_{A'} A'(d)$ is tangentially flat for every d. Since

$$f' \otimes_{A'} A'(d) = f \otimes_A A(d) \otimes_{A(d)} A'(d)$$

and $f \otimes_A A(d)$ is tangentially flat by 3.8, we may assume that A and A' are Artinian local rings. Let $x = (x_1, \ldots, x_N) \in A'$ and $q = (q_1, \ldots, q_N) \in \mathbb{N}$ be N-tuples such that all coordinates of q are positive and such that for every $d \in \mathbb{N}$ the set of x_i's with $q_i = d$ form a generating set of the ideal $F_{A'}^d$. Note that it is possible to find such N-tuples since A' is Artinian. By construction, $F_{A'}$ is the q-adic filtration over $F_A A'$ defined by x, i.e.,

$$F_{A'}^d = \sum_{i+<q,j>\geq d} F_A^i \cdot x^j A' \quad \text{for every } d \in \mathbb{N}.$$

Similarly, $F_{B'}$ is q-adic over $F_B B'$,

$$F_{B'}^d = \sum_{i+<q,j>\geq d} F_B^i \cdot x^j B' \quad \text{for every } d \in \mathbb{N}.$$

Let $X = (X_1, \ldots, X_N)$ be an N-tuple of indeterminates and equip the power series rings

$$A'' := A[[X]] \quad \text{and} \quad B'' := B[[X]]$$

with the q-adic filtrations defined by X over $F_A A''$ and $F_B B''$, respectively. Then, as a filtered ring, A' is the factor ring of A'' modulo the ideal generated by the linear polynomials $X_i - x_i$, and similarly, $B' = B''/(X_i - x_i | i = 1, \ldots, N)$. By 3.8 it is sufficient to show that B'' is tangentially flat over A''. However, this is a consequence of 3.9.

4. Relation to flatness

Remark

In this chapter we prove that tangential flatness implies flatness. Moreover we prove theorems saying that tangential flatness is equivalent to flatness plus 'something else'. So we will prove that $f : (A, m) \to (B, n)$ is tangentially flat if and only if so is $f \otimes A/I$, f is flat, and $\mathrm{in}(IB) = \mathrm{in}(I)\,\mathrm{G}(B)$. Another theorem says that $A \to B = A[[X_1, \ldots, X_N]]/(f_1, \ldots, f_n)$ is tangentially flat if and only if it is flat and the orders of the generators f_i are preserved in passing to the residue classes modulo $mA[[X_1, \ldots, X_N]]$. In fact the latter result is the most important tool to construct examples of tangentially flat homomorphisms. As a preparation we describe the completions of free modules.

4.1. Completion of free modules

Let (A, m) be a filtered local ring and $X = \oplus_{\lambda \in \Lambda} A$ a free A-module. Equip X with the filtration F_X such that

$$F_X^d := F_A^d \cdot X \quad \text{for every } d.$$

Then

(i) The completion of X can be identified with a submodule of a direct product as follows.

$$(X, F_X)^\wedge = \left\{ (a_\lambda)_{\lambda \in \Lambda} \in \prod_{\lambda \in \Lambda} (A, F_A)^\wedge \,\middle|\, \begin{array}{l}\text{given } d \in \mathbb{N}, \text{ there are at most} \\ \text{finitely many } \lambda \text{ with } a_\lambda \notin F_{A^\wedge}^d\end{array}\right\}$$

(ii) $(X, F_X)^\wedge$ is flat over $(A, F_A)^\wedge$.

(iii) For every ideal I of A, $(X/IX)^\wedge \cong X^\wedge/IX^\wedge$.

Proof. Form the direct sum analogous to X with A replaced by its completion (with respect to F_A),

$$X^* = \bigoplus_{\lambda \in \Lambda} A^\wedge.$$

By 1.18,

$$A^\wedge/F_A^d A^\wedge \cong (A/F_A^d)^\wedge \cong A/F_A^d$$

where the second isomorphism results from the fact that the topology of A/F_A^d is discrete. Therefore, the map $X \to X^*$ induced by the natural homomorphism $A \to A^\wedge$ defines bijections

$$X/F_A^d X \to X^*/F_A^d X^*,$$

and $X^\wedge$ may be identified with the limit of the modules $X^*/F_A^d X^*$. Hence, in proving the above statements, we may assume that A is complete.

(i). Let Y be the module on the left hand side of (i), i.e.,

$$Y = \left\{ (a_\lambda)_{\lambda \in \Lambda} \in \prod_{\lambda \in \Lambda} A \;\middle|\; \text{given } d \in \mathbb{N}, \, a_\lambda \notin F_A^d \text{ for at most finitely many } \lambda \right\}$$

Given an element $(a_\lambda)_{\lambda \in \Lambda} \in Y$, there are only finitely many a_λ which are non-zero modulo F_A^{d+1}. So there are well defined maps

$$f_d : Y \to X/F_A^{d+1} X = X/F_X^{d+1}, \quad (a_\lambda)_{\lambda \in \Lambda} \mapsto (a_\lambda \bmod F_A^{d+1})_{\lambda \in \Lambda},$$

which are compatible with the natural projections $X/F_X^{d+1} \to X/F_X^d$ and hence define a homomorphism

$$f : Y \to X^\wedge, \quad y := (a_\lambda)_{\lambda \in \Lambda} \mapsto (f_d(y))_{d \in \mathbb{N}}$$

We want to prove that this latter homomorphism f is bijective. To do so, consider the map

$$g : X^\wedge \to Y, \quad x := (x_d \bmod F_X^{d+1}) \mapsto g(x)$$

which is defined as follows. For every d write $x_d = (a_{d,\lambda})_{\lambda \in \Lambda}$ with $a_{d,\lambda} \in A$. Since $x \in X^\wedge$, the coordinates of x satisfy $x_d - x_{d+1} \in F_X^{d+1}$, hence

$$a_{d,\lambda} - a_{d+1,\lambda} \in F_A^{d+1} \quad \text{for every } d \text{ and every } \lambda.$$

In other words, the family $(a_{d,\lambda} \bmod F_A^{d+1})_{d \in \mathbb{N}}$ is in the completion of A. Since A is assumed to be complete, this family is in the image of the canonical map $A \to A^\wedge$. There is a uniquely determined $a_\lambda \in A$ such that $a_\lambda - a_{d,\lambda} \in F_A^{d+1}$ for every d. Define $y' := (a_\lambda)_{\lambda \in \Lambda}$ to be the image of x under g. By construction,

$$(1) \qquad\qquad y' - x_d \in F_X^{d+1},$$

where x_d, being an element of the direct sum X, has only finitely many non-zero coordinates. Hence all but finitely many coordinates of y' are in F_X^{d+1} (d arbitrary), and y' is a well-defined element of Y. This shows the above definition of g is correct.

Next consider the image of x under the composition $f \circ g$. In view of (1),

$$f \circ g(x) = f(y') = (f_d(y'))_{d \in \mathbb{N}} = (y' \bmod F_X^{d+1})_{d \in \mathbb{N}} = (x_d \bmod F_X^{d+1})_{d \in \mathbb{N}} = x$$

i.e., $f \circ g$ is the identity map. In particular, f is surjective.

To prove that f is injective, note that $\mathrm{Ker}(f_d)$ consists of those $y := (a_\lambda)_{\lambda \in \Lambda} \in Y$ which have all coordinates a_λ in F_A^{d+1}. If $y := (a_\lambda)_{\lambda \in \Lambda} \in Y$ is in the kernel of

f, then $y \in \mathrm{Ker}(f_d)$ for every d, hence all coordinates of y are in $\underset{d \in \mathbb{N}}{\cap} F_A^d = 0$.

This completes the proof the assertion (i).

(ii). As above we may assume that A is complete. Let $I \subseteq A$ be an ideal. We have to show, the natural map

$$f : I \otimes_A X^\wedge \to X^\wedge, \quad i \otimes x \mapsto i \cdot x,$$

is injective ([Ma86], Th. 7.7). Let $i_j \in I$ and $x_j \in X^\wedge$ be elements with $f(\sum_{j=1}^r i_j \otimes x_j) = 0$. We have to show, $\sum_{j=1}^r i_j \otimes x_j = 0$. By assumption

$$(2) \qquad \sum_{j=1}^r i_j x_j = 0.$$

Write $x_j = (x_{j,\lambda})_{\lambda \in \Lambda} \in \prod_{\lambda \in \Lambda} A$. Then, given $d \in \mathbb{N}$,

$$x_{j,\lambda} \in (F_A^1)^d \text{ except for finitely many } \lambda.$$

We will use the following notation.

$$x_{*,\lambda} := (x_{1,\lambda}, \dots, x_{r,\lambda}) \in A^r,$$

$$L := \sum_{\lambda \in \Lambda} x_{*,\lambda} \cdot A \quad (\subseteq A^r),$$

$$\mathrm{d}(\lambda) := \sup\{d \in \mathbb{N} | x_{*,\lambda} \in (F_A^1)^d \cdot A^r\}$$

Since A is Noetherian, L is generated by finitely many of the elements $x_{*,\lambda}$, say

$$L := x_{*,\lambda_1} A + \dots + x_{*,\lambda_s} A$$

Let c be the constant of the Artin-Rees lemma ([Ma86], Th. 8.5), i.e., such that

$$((F_A^1)^{d+c} \cdot A^r) \cap L \subseteq (F_A^1)^d \cdot L \text{ for every } d \in \mathbb{N}.$$

Then, if $(F_A^1)^\infty$ denotes the zero ideal, $x_{*,\lambda} \in (F_A^1)^{\mathrm{d}(\lambda)-c} \cdot L$, hence

$$(3) \qquad x_{*,\lambda} = \sum_{k=1}^s a_{\lambda,k} \cdot x_{*,\lambda_k} \text{ with } a_{\lambda,k} \in (F_A^1)^{\mathrm{d}(\lambda)-c}$$

for every λ and every j. Since $x_j = (x_{j,\lambda})_{\lambda \in \Lambda}$ is in $X^\wedge$, for every given $d \in \mathbb{N}$, there are only finitely many $\lambda \in \Lambda$ such that $\mathrm{d}(\lambda) < d$. Therefore,

$$a_k := (a_{\lambda,k})_{\lambda \in \Lambda} \text{ is in } X^\wedge \text{ for every } k.$$

Moreover, the identities (3) and (2) imply

$$x_j = \sum_{k=1}^{s} x_{j,\lambda_k} \cdot a_k \text{ for every } j$$

$$0 = \sum_{j=1}^{r} i_j \cdot x_{j,\lambda_k} \text{ for every } k.$$

hence $\sum_{j=1}^{r} i_j \otimes x_j = \sum_{k=1}^{s} \sum_{j=1}^{r} i_j x_{j,\lambda_k} \otimes a_k = 0$, as required.

(iii). By 1.18 it is sufficient to show that $F_X = F_A X$ is an Artin-Rees filtration. But this is the case, since F_A has the Artin-Rees property and X is a free module.

4.2. Tangential flatness implies flatness

Let $f : (A, m) \to (B, n)$ be a homomorphism of filtered local rings. Assume that f is tangentially flat. Then

(i) $G(B)$ is free as a module over $G(A)$,

(ii) B is flat over A.

Proof. (i). Choose a family $(b_\lambda)_{\lambda \in \Lambda}$ of elements from B such that the initial elements

$$\text{in}(b_\lambda) \in G(B)$$

represent a homogeneous vector space basis of $G(B)/\text{in}(m) G(B)$ over the field $G(A)/\text{in}(m)(\cong G(A/m) \cong A/m)$. Since $G(B)$ is $G(A)$-flat, the elements $\text{in}(b_\lambda)$ are linearly independent over $G(A)$ by 2.7 and the second remark of 3.11. Further,

$$G(B) = \sum_{\lambda \in \Lambda} G(A) \cdot \text{in}(b_\lambda) + \text{in}(m) G(B).$$

hence, by the graded Nakayama lemma 2.1, the initial forms $\text{in}(b_\lambda)$ generate $G(B)$ as a $G(A)$-module (cf the first remark of 3.11). We have proved assertion (i).

(ii). The fact that the initial forms of the b_λ's form a free generating set over $G(A)$ means that the canonical homomorphisms

$$\bigoplus_{\lambda \in \Lambda} A/F_A^{d-d_\lambda} \to B/F_B^d, \quad (a_\lambda \bmod F_A^{d-d_\lambda})_{\lambda \in \Lambda} \mapsto \sum_{\lambda \in \Lambda} a_\lambda b_\lambda \bmod F_B^d,$$

are even isomorphisms for every $d \in \mathbb{N}$ (see 2.6). Here $d_\lambda := \text{ord}_{F_B} b_\lambda$. These homomorphisms form a morphism of inverse systems. Taking the inverse limit, we obtain

$$(B, F_B)^\wedge := \varprojlim_{d \in \mathbb{N}} B/F_B^d \cong \varprojlim_{d \in \mathbb{N}} \bigoplus_{\lambda \in \Lambda} A/F_A^{d-d_\lambda}$$

(1)
$$= \varprojlim_{d \in \mathbb{N}} \prod_{n=1}^{\infty} \bigoplus_{d_\lambda = n} A/F_A^{d-n}$$

$$\cong \prod_{n=1}^{\infty} \varprojlim_{d \in \mathbb{N}} \bigoplus_{d_\lambda = n} A/F_A^{d-n}$$

$$= \prod_{n=1}^{\infty} (\bigoplus_{d_\lambda = n} A)^\wedge$$

Here "$\wedge$" denotes the completion with respect to the m-adic topology (see 1.15(iv) and (vi)). Identity (1) follows from the fact that for every fixed d there are only finitely many n such that the module under the product sign is non-zero. Since the modules

$$(\bigoplus_{d_\lambda = n} A)^\wedge$$

are flat over $A^\wedge$ by 4.1(ii), the claim follows. Note that the direct product of flat modules over a Noetherian ring is flat ([Ma86], Ex. 7.4).

4.3. Tangential flatness and initial ideals I

Let $f : (A, m) \to (B, n)$ be a homomorphism of filtered local rings. Suppose the following conditions are satisfied.

1. f is flat.
2. $\mathrm{in}(IB) = \mathrm{in}(I)\,G(B)$ for every proper ideal $I \subset A$.

Then f is tangentially flat.

Proof. By 3.15 it will be sufficient to show that the following conditions are satisfied.

(1) $G(B_0)$ is A_0-flat, where $A_0 := A/F_A^1$ and $B_0 := B/F_A^1 B$.

(2) The canonical surjections

$$G(A)(i) \otimes_A B(d-i) \to G_{F_A}(B)(i) \otimes_B B(d-i)$$

are bijective.

(3) The canonical surjections

$$G_{F_A}(B)(i) \otimes_B B(d-i) \to G_{F_A}(B(d))(i)$$

are bijective.

Consider the exact sequences

$$0 \to G(B_0)(d) \to B_0(d) \to B_0(d-1) \to 0.$$

In order to prove (1), it will be sufficient to show that $B_0(d)$ is A_0-flat for every d, i.e.,

$$\mathrm{Tor}_1^{A_0}(B_0(d-1), A/I) = 0$$

for every ideal $I \subseteq A$ containing F_A^1 and every d. Calculate the Tor-module, using the exact sequence

$$0 \to F_{B_0}^d \xrightarrow{\alpha} B_0 \to B_0(d-1) \to 0.$$

Since B_0 is flat over A_0, it will be sufficient to show that $\alpha \otimes_A A/I$ is injective for every ideal $I \subseteq A$ containing F_A^1. Note that

$$F_{B_0}^d = F_B^d \cdot B_0 = (F_B^d + F_A^1 B)/F_A^1 B.$$

So we have to prove, the canonical homomorphism

$$(F_B^d + F_A^1 B)/(IF_B^d + F_A^1 B) \to B/(IB + F_A^1 B),$$

is injective, i.e., we must show

$$F_B^d \cap (IB + F_A^1 B) \subseteq IF_B^d + F_A^1 B.$$

But by 3.13 this is a consequence of the identities $\mathrm{in}(IB + F_A^1 B) = \mathrm{in}(I + F_A^1) G(B)$. We have proved condition (1) is satisfied. Condition (2) follows directly from the fact that B is A-flat: to get the required bijection, apply the functor $\otimes_B B(d-i)$ to the canonical homomorphism

$$G(A) \otimes_A B \to G_{F_A}(B)$$

which is bijective according to the local flatness criterion ([Ma86], Th. 22.3(4')). We have yet to show condition (3) is satisfied, i.e., the canonical surjections

$$G_{F_A}(B)(i) \otimes_B B(d-i) \to G_{F_A}(B(d))(i)$$

are bijective. In view of 3.14 this is equivalent to the identities $\mathrm{in}(F_A^i B) = \mathrm{in}(F_A^i) G(B)$. So the claim follows since, by assumption, $\mathrm{in}(IB) = \mathrm{in}(I) G(B)$ for every proper ideal I of A.

4.4. Tangential flatness and initial ideals II

Let $f(A, m) \to (B, n)$ be a flat homomorphism of filtered local rings and $I \subset A$ a proper ideal. Then the following conditions are equivalent.

(i) f is tangentially flat.

(ii) $f \otimes_A A/I$ is tangentially flat and $\text{in}(IB) = \text{in}(I)\,G(B)$.

Proof. (i)$\Rightarrow$(ii). This was already proved (see 3.8).

(ii)$\Rightarrow$(i). Step 1: We prove that, for every proper ideal $J \subset A$, the functor $\otimes_A A/J$ preserves condition (ii).

Indicate by a prime passage to the residue classes modulo J, i.e. write

$$A' := A/J, \quad B' := B/JB, \quad I' := I \cdot A', \quad \text{and} \quad f' := f \otimes_A A/J : A' \to B'.$$

We have to show that $f' \otimes_{A'} A'/I'$ is tangentially flat and that $\text{in}(I'B') = \text{in}(I')\,G(B')$. The first assertion follows from the fact that

$$f' \otimes_{A'} A'/I' = f \otimes_A A/(I + J) = (f \otimes_A A/I) \otimes_A A/J$$

(by assumption (ii) and 3.8). As for the second, note that since $f \otimes_A A/I$ is tangentially flat,

$$\begin{array}{ccc} \text{in}(J \cdot B/IB) & = & \text{in}(J \cdot A/I)\,G(B/IB) \\ \| & & \| \\ \text{in}(IB + JB/IB) & & \text{in}(I + J/I)\,G(B/IB) \end{array}$$

hence by 1.12(4),

$$\text{in}(IB + JB) = \text{in}(I + J)\,G(B) + \text{in}(IB).$$

By assumption (ii), $\text{in}(IB) = \text{in}(I)\,G(B)$, so the second term on the right may be omitted. Therefore, $\text{in}(IB + JB) = \text{in}(I + J)\,G(B)$ hence $\text{in}(I \cdot B') = \text{in}(I \cdot A')\,G(B')$ as required.

Step 2: Let S be the set of proper ideals $J \subset A$ such that $f \otimes_A A/J$ is not tangentially flat. We want to prove that S is empty. Suppose it is not. Then, since A is Noetherian, S contains some J which is maximal in S with respect to inclusion. Replace f by $f \otimes_A A/J$. According to the first step, assumption (ii) is preserved under this operation. So we may assume that S contains the zero ideal as its only element. We shall prove now that f is tangentially flat. This contradiction will complete the proof. To prove tangential flatness of f, it will be sufficient (by 4.3) to show that $\text{in}(JB) = \text{in}(J)\,G(B)$ for every ideal $J \subset A$. In case J is the zero ideal, this is trivial. So assume $J \neq (0)$. Choose any non-zero element

$$0 \neq t \in J.$$

Since tA is not in S, $f \otimes A/tA$ is tangentially flat, hence

$$\text{in}(JB/tB) = \text{in}(J/tA)\,G(B/tB),$$

i.e., $\text{in}(JB) = \text{in}(J)\,G(B) + \text{in}(tB)$. It will be sufficient to show

(1) $$\text{in}(tB) \subseteq \text{in}(tA)\,G(B),$$

for, this implies $\mathrm{in}(tB) \subseteq \mathrm{in}(J)\,G(B)$ hence $\mathrm{in}(JB) = \mathrm{in}(J)\,G(B)$. In order to prove inclusion (1), consider the commutative diagram of natural homomorphisms,

$$
\begin{array}{ccc}
G(B) & \overset{\alpha}{\longrightarrow} & G(B/IB) \\
\downarrow \delta & & \beta \downarrow \\
G(B/tB) & \overset{\gamma}{\longrightarrow} & G(B/(IB + tB))
\end{array}
$$

Obviously,

$$
\mathrm{in}(tB) = \mathrm{Ker}(\delta) \subseteq \mathrm{Ker}(\gamma\delta) = \mathrm{Ker}(\beta\alpha).
$$

On the other hand, since $f \otimes A/I$ is tangentially flat,

$$
\mathrm{Ker}(\beta) = \mathrm{in}(t \cdot B/IB) = \mathrm{in}(t \cdot A/I)\,G(B/IB) = \mathrm{in}(I + tA) \cdot G(B/IB),
$$

hence $\mathrm{in}(tB) \subseteq \mathrm{in}(I + tA)\,G(B) + \mathrm{in}(IB) = \mathrm{in}(I + tA)\,G(B)$ (we are using our assumption that $\mathrm{in}(IB) = \mathrm{in}(I)\,G(B)$). Therefore,

$$
\mathrm{in}(tB) = \mathrm{in}(I + tA)\,G(B) \cap \mathrm{in}(tB).
$$

To calculate the intersection on the right, consider the short exact sequence

$$
0 \;\longrightarrow\; \mathrm{in}(tB)/\mathrm{in}(tA)\,G(B) \;\overset{\varepsilon}{\longrightarrow}\; G(B)/\mathrm{in}(tA)\,G(B) \;\longrightarrow\; G(B/tB) \;\longrightarrow\; 0.
$$

Since $f \otimes A/tA$ is tangentially flat, the exactness of this sequence is preserved when it is tensored with $G(A)/\mathrm{in}(I + tA)$ over $G(A/tA)$. In particular, the tensor product $\bar{\varepsilon} := \varepsilon \otimes G(A)/\mathrm{in}(I + tA)$ is injective,

$$
\bar{\varepsilon} : \mathrm{in}(tB)/(\mathrm{in}(tA)\,G(B) + \mathrm{in}(I + tA)\,\mathrm{in}(tB)) \rightarrow G(B)/\mathrm{in}(I + tA)\,G(B),
$$

i.e.,

$$
\mathrm{in}(tB) = \mathrm{in}(I + tA)\,G(B) \cap \mathrm{in}(tB) = \mathrm{in}(tA)\,G(B) + \mathrm{in}(I + tA)\,\mathrm{in}(tB).
$$

Since the homogeneous parts of $G(B)$ are modules of finite length over $G(B)(0)$, the graded Nakayama lemma 2.1 implies

$$
\mathrm{in}(tB) = \mathrm{in}(tA)\,G(B).
$$

4.5. Initial forms of extension ideals

Let $f : (A, m) \rightarrow (B, n)$ be a homomorphism of filtered local rings and let $I \subseteq A$, $J \subseteq B$ be ideals. Assume that f is tangentially flat. Then the following conditions are equivalent.

(i) $\mathrm{in}(I \cdot B/J) = \mathrm{in}(I)\,G(B/J)$

(ii) The canonical map $\mathrm{in}(J) \rightarrow \mathrm{in}(J \bmod IB)$ is surjective.

Proof. Consider the following commutative diagram of natural homomorphisms with exact rows and columns.

$$
\begin{array}{ccccccccc}
 & & 0 & & 0 & & 0 & & \\
 & & \downarrow & & \downarrow & & \downarrow & & \\
0 & \to & \text{in}(J) \cap \text{in}(IB) & \to & \text{in}(IB) & \to & \text{in}(I \cdot B/J) & & \\
 & & \downarrow & & \downarrow & & \downarrow & & \\
0 & \to & \text{in}(J) & \to & G(B) & \overset{\phi}{\to} & G(B/J) & \to & 0 \\
 & & \downarrow & & \downarrow & & \downarrow & & \\
0 & \to & \text{in}(J \bmod IB) & \to & G(B/IB) & \to & G(B/(J+IB)) & \to & 0 \\
 & & & & \downarrow & & \downarrow & & \\
 & & & & 0 & & 0 & &
\end{array}
$$

By 3.8, since f is tangentially flat,

$$\phi(\text{in}(IB)) = \phi(\text{in}(I) \cdot G(B)) = \text{in}(I) \cdot \phi(G(B)) = \text{in}(I) \cdot G(B/J)$$

So condition (i) means that the first row of the diagram is short exact. By the nine lemma this is equivalent to short exactness of the left column, which is just the claim of (ii).

Remark

Later we will see that a rather large class of flat local homomorphisms of filtered complete local rings factors into a tangentially flat one and a surjection (see 5.9). So it is interesting to know what is the condition upon the ideal in a tangentially flat extension to give a tangentially flat factor ring. The theorem that follows gives an answer. It will be typically applied with I equal to the maximal ideal of A and B a power series ring over A (cf the second part of 3.9).

4.6. Tangential flatness and liftable standard bases

Let $f : (A, m) \to (B, n)$ be a homomorphism of filtered local rings and let $I \subseteq A$, $J \subseteq B$ be proper ideals. Assume that

  1. B is tangentially flat over A
  2. $B^* := B/J$ is flat over A.

Then the following are equivalent.

  (i) B^* is tangentially flat over A.
  (ii) B^*/IB^* is tangentially flat over A/I and the canonical map

$$\text{in}(J) \to \text{in}(J \cdot B/IB)$$

is surjective.

(iii) B^*/IB^* is tangentially flat over A/I and for every element

$$b' \in J \cdot B/IB$$

there is a lift $b \in J$ with $\operatorname{ord}_B b = \operatorname{ord}_{B/IB} b'$.

(iv) B^*/IB^* is tangentially flat over A/I and there is a standard basis $f' = (f'_\lambda)_{\lambda \in \Lambda}$ of $J \cdot B/IB$ which can be lifted to a family $f = (f_\lambda)_{\lambda \in \Lambda}$ of elements from J preserving the orders,

$$\operatorname{ord}_B f_\lambda = \operatorname{ord}_{B/IB} f'_\lambda \text{ for every } \lambda$$

(in which case every such standard base can be lifted this way and every such lift is automatically a standard base of J).

Proof. (i) $\Leftrightarrow$ (ii). By 4.4, tangential flatness of $A \to B/J$ is equivalent to tangential flatness of $(A \to B/J) \otimes A/I = A/I \to B/(J + IB)$ together with the identity $\operatorname{in}(I \cdot B/J) = \operatorname{in}(I) \operatorname{G}(B/J)$. Since f is supposed to be tangentially flat, the latter identity in turn is equivalent to the surjectivity condition of (ii) (see 4.5).

(ii)$\Rightarrow$(iii). Surjectivity of the canonical map $\operatorname{in}(J) \to \operatorname{in}(J \cdot B/IB)$ implies

$$(J + IB) \cap F_B^d \subseteq (J + IB) \cap F_B^{d+1} + J \cap F_B^d + IB.$$

Iterating this inclusion, the first term on the right can be replaced by $(J+IB) \cap F_B^k$ with k arbitrarily large. Since F_B is an Artin-Rees filtration, we may assume, the latter intersection is contained in $(J + IB) \cdot F_B^d$ so that it can be omitted:

$$(J + IB) \cap F_B^d \subseteq J \cap F_B^d + IB.$$

The last inclusion has the following interpretation. Given an element $b' \in J \cdot B/IB$ of finite order d, there is an element $b \in J$ of order d with residue class modulo IB equal to b'. But this is the claim of (iii).

(iii)$\Rightarrow$(iv). Trivial.

(iv)$\Rightarrow$(ii). The canonical $\operatorname{G}(B)$-module homomorphism

$$\operatorname{in}(J) \to \operatorname{in}(J \cdot B/IB)$$

maps the family $(\operatorname{in}(f_\lambda))_{\lambda \in \Lambda}$ into a system of generators $(\operatorname{in}(f'_\lambda))_{\lambda \in \Lambda}$ of $\operatorname{in}(J \cdot B/IB)$, hence it is surjective.

We have yet to prove the statement of condition (iv) enclosed in parentheses. From the above proof it is clear that every standard basis can be lifted in the required way provided this is possible for at least one. Let $f = (f_\lambda)_{\lambda \in \Lambda}$ be a lift of a standard basis $f' = (f'_\lambda)_{\lambda \in \Lambda}$ of $J \cdot B/IB$ satisfying the conditions of (iv). We have to show that f is a standard basis of J, i.e., the family $\operatorname{in}(f) := (\operatorname{in}(f_\lambda))_{\lambda \in \Lambda}$ generates $\operatorname{in}(J)$. By construction,

$$\operatorname{in}(J) \subseteq \operatorname{in}(J + IB) = (\operatorname{in}(f)) + \operatorname{in}(IB),$$

hence

$$\mathrm{in}(J) = (\mathrm{in}(f)) + \mathrm{in}(IB) \cap \mathrm{in}(J) = (\mathrm{in}(f)) + \mathrm{in}(I)\,\mathrm{G}(B) \cap \mathrm{in}(J).$$

Note that, since B is tangentially flat over A, $\mathrm{in}(IB) = \mathrm{in}(I)\,\mathrm{G}(B)$. Tangential flatness of B/J over A implies that the homomorphism

$$(\mathrm{in}(J) \rightarrow \mathrm{G}(B)) \otimes_{\mathrm{G}(A)} \mathrm{G}(A)/\mathrm{in}(I)$$

induced by the canonical embedding $\mathrm{in}(J) \rightarrow \mathrm{G}(B)$ is injective. So $\mathrm{in}(I)\,\mathrm{G}(B) \cap \mathrm{in}(J) = \mathrm{in}(I)\,\mathrm{in}(J)$ and the above identity writes

$$\mathrm{in}(J) = (\mathrm{in}(f)) + \mathrm{in}(I)\,\mathrm{in}(J).$$

The graded Nakayama lemma 2.1 implies $\mathrm{in}(J) = (\mathrm{in}(f))$, i.e., f is a standard basis of J as claimed.

5. Distinguished bases

Remark

In this chapter we will introduce the notion of distinguished base for a homomorphism of filtered local rings $f : (A, m) \rightarrow (B, n)$. In case f is flat this is a substitute for a free set of generators of B over A taking into account the filtrations. In the complete case we will use distinguished bases to define structure constants of B over A and characterize tangential flatness in terms of these structure constants. Further we will prove a version of Cohen factorization theorem, and relate tangential flatness to the behavior of the order function under flat local extensions.

5.1. Definition

Let $f : (A, m) \rightarrow (B, n)$ be a local homomorphism of local rings and $\bar{F}$ a filtration on the special fibre $\bar{B} = B/mB$ of f. An $\bar{F}$-*distinguished basis* of B over A is a pair (b, d) of maps

$$b : \Lambda \rightarrow B \text{ and } d : \Lambda \rightarrow \mathbb{N}$$

defined on some set Λ and satisfying the following conditions.

1. The initial forms $in(\overline{b(\lambda)})$ of the residue classes

$$\overline{b(\lambda)} := b(\lambda) \bmod mB$$

   form a vector space basis of $G_{\bar{F}}(\bar{B})$ over A/m.
2. $d(\lambda) = ord_{\bar{F}}(\overline{b(\lambda)})$ for every $\lambda \in \Lambda$.
3. For every $d \in \mathbb{N}$ there is some $k(d) \in \mathbb{N}$ such that $d(\lambda) \geq k(d)$ implies $b(\lambda) \in n^d$.

Now let F be a filtration on B and $\bar{F}$ the filtration induced by F on $\bar{B}$. Then an F-*distinguished basis* of B over A is an $\bar{F}$-distinguished basis (b, d) of B over A such that $d(\lambda) = ord_F b(\lambda)$ for every $\lambda \in \Lambda$. Let $f : (A, m) \rightarrow (B, n)$ be a homomorphism of filtered local rings. Then a *distinguished basis* for f is by definition an F_B-distinguished basis of B over A.

Remark

The existence of distinguished bases is obvious from the definition, and it is quite easy to see that the operation $f \mapsto f \otimes A/I$ preserves distinguished bases. We state this for reference below. Later we will clarify in which sense a distinguished basis can be considered to be a basis of B over A.

5.2. Existence and stability under base change

Let $f : (A, m) \to (B, n)$ be a local homomorphism of local rings. Then

(i) For every ring filtration $\bar{F}$ on the special fibre $\bar{B} := B/mB$ there is an $\bar{F}$-distinguished basis of B over A.

(ii) For every ring filtration F on B there is an F-distinguished basis of B over A.

(iii) Let F be a ring filtration on B, (b, d) an F-distinguished of B over A, and $I \subset A$ a proper ideal. Define

$$A' := A/I, \quad B' := B/IB, \quad b'(\lambda) := (b(\lambda) \bmod IB).$$

Then (b', d) is an F'-distinguished basis of B' over A' where F' is the filtration on B' induced by F.

Proof. (i) and **(ii)**. Choose elements $\bar{b}_\lambda \in \bar{B}$ such that the family $(in(\bar{b}_\lambda))_{\lambda \in \Lambda}$ of initial forms with respect to $\bar{F}$ is a (homogeneous) vector space basis of $G(\bar{B})$ over A/m. Since $\bar{F}$ is an Artin-Rees filtration, there is, for every $d \in \mathbb{N}$ some $k \in \mathbb{N}$ such that $\bar{b}_\lambda \in \bar{F}^k$ implies $\bar{b}_\lambda \in n^d \bar{B}$ (see 1.14(iii)). For every $\lambda \in \Lambda$ let $b(\lambda) \in B$ be any representative of $\bar{b}_\lambda$ having the same order with respect to the maximal ideal like $\bar{b}_\lambda$ and define $d(\lambda) := \text{ord}_{\bar{F}}\,\bar{b}_\lambda$. Then the pair (b, d) is an $\bar{F}$-distinguished basis of B over A. This proves (i). If the representative $b(\lambda) \in B$ of $\bar{b}_\lambda$ is chosen such that $\text{ord}_F\,b(\lambda) = \text{ord}_{\bar{F}}\,\bar{b}_\lambda$ instead, then (b, d) is even F-distinguished.

(iii). Let $\bar{F}$ be the filtration on $\bar{B}$ induced by F. By assumption (b, d) is $\bar{F}$-distinguished for B over A hence so is (b', d) for B' over A'. All we have to show is that $\text{ord}\,b'(\lambda) = d(\lambda)$ for every λ. However, this is obvious from the inequalities,

$$d(\lambda) = \text{ord}\,b(\lambda) \leq \text{ord}\,b'(\lambda) \leq \text{ord}\,\bar{b}_\lambda = d(\lambda).$$

5.3. Distinguished bases and free generators for the graded algebra

Let $f : (A, m) \to (B, n)$ be a homomorphism of filtered local rings and (b, d) a distinguished basis for f. Assume that f is tangentially flat. Then the initial elements

$$in(b(\lambda)) = (b(\lambda) \bmod F_B^{d(\lambda)+1}) \in G(B)$$

form a free generating system of $G(B)$ over $G(A)$.

Proof. From the definition of an F_B-distinguished basis we see that, for every $d \in \mathbb{N}$,

$$F_B^d \subseteq \sum_{\lambda \in \Lambda} A \cdot b(\lambda) + F_B^{d+1} + mB.$$

Therefore

$$B \subseteq \sum_{\lambda \in \Lambda} A \cdot b(\lambda) + F_B^d + mB$$

for $d = 0,1,2,\ldots$. Iterating the latter inclusion, the maximal ideal m of A may be replaced by an arbitrary power m^i. Since F_B is cofinite, $m^i B \subseteq F_B^d$ for large i, i.e.,

$$B \subseteq \sum_{\lambda \in \Lambda} A \cdot b(\lambda) + F_B^d$$

for every $d \in \mathbb{N}$. By 2.6 it will be sufficient to show that, for every $d \in \mathbb{N}$,

$$\sum_{\lambda \in \Lambda} a_\lambda\, b(\lambda) \in F_B^d \text{ implies } a_\lambda \in F_A^{d-d(\lambda)}.$$

The latter means that the initial forms $\text{in}(b(\lambda)) \in G(B)$ are linearly independent over $G(A)$. Since (b, d) is F_B-distinguished, the images under the canonical surjection

$$G(B) \to G(B)/\text{in}(m)\, G(B) \cong G(B/mB)$$

of the $\text{in}(b(\lambda))$'s form a vector space basis of $G(B/mB)$ over the field $G(A/m) \cong G(A)/\text{in}(m)$. Hence the claim follows from 2.7 (see also the second remark of 3.11).

5.4. Minimally lifted filtrations

Let $f : (A, m) \to (B, n)$ be a homomorphism of filtered local rings, and (b, d) a distinguished basis for f. Then

$$(1) \qquad F_B'^d := \sum_{(i,\lambda) \in \mathbb{N}^n \times \Lambda^n} F_A^{d-<i,d(\lambda)>} \cdot B \cdot b(\lambda)^i \subseteq F_B^d$$

for every $d \in \mathbb{N}$ as is easily seen from the fact that $b(\lambda)$ has F_B-order $d(\lambda)$ for every λ (see 5.1). Here the sum is taken over the non-negative integers n and all pairs $(i, \lambda) \in \mathbb{N}^n \times \Lambda^n$. Moreover,

$$d(\lambda) := (d(\lambda_1), \ldots, d(\lambda_n))$$
$$< i, d(\lambda) > := i_1 \cdot d(\lambda_1) + \ldots + i_n \cdot d(\lambda_n)$$
$$b(\lambda)^i := b(\lambda_1)^{i_1} \cdot \ldots \cdot b(\lambda_n)^{i_n}$$

for $i = (i_1, \ldots, i_n)$ and $\lambda = (\lambda_1, \ldots, \lambda_n)$. Passing to the residue classes modulo mB, the above inclusion gives an identity, for, by definition 5.1,

$$F_B^d \cdot \bar{B} = \sum_{d(\lambda) \geq d} (A/m) \cdot \overline{b(\lambda)} + F_B^{d+1} \cdot \bar{B} \subseteq F_B'^d \cdot \bar{B} + F_B^{d+1} \cdot \bar{B} (\subseteq F_B^d \cdot \bar{B}).$$

Iterating the identity $F_B^d \cdot \bar{B} = F_B'^d \cdot \bar{B} + F_B^{d+1} \cdot \bar{B}$, the last term $F_B^{d+1} \cdot \bar{B}$ on the right can be replaced by $F_B^{d+k} \cdot \bar{B}$ with k arbitrarily large, hence by $n \cdot F_B^d \cdot \bar{B}$, since F_B is an Artin-Rees filtration. The Nakayama lemma implies that the term can be omitted at all, i.e., F_B and F_B' induce one and the same filtration on the special fibre $\bar{B} = B/mB$. So (1) defines a lift to B of the filtration $F_{\bar{B}} := F_B \cdot \bar{B}$ induced by F_B on the special fibre $\bar{B}$. By construction, $F' := (F_B'^d)_{d \in \mathbb{N}}$ is a ring filtration of B (cf 1.11) such that $f : (A, F_A) \to (B, F')$ is a homomorphism of filtered local rings. Moreover, it is a lower bound for the possible filtrations F_B making f a homomorphism of filtered local rings and inducing the given filtration $F_{\bar{B}}$ on the special fibre.

Given a homomorphism $f : (A, m) \to (B, n)$ of filtered local rings, we will say that F_B is F_A-*minimal*, if there is an $F_B \cdot B/mB$-distinguished basis (b, d) of B over A with

$$F_B'^d = \sum_{(i, \lambda) \in \mathbb{N}^n \times \Lambda^n} F_A^{d - <i, d(\lambda)>} \cdot B \cdot b(\lambda)^i \text{ for every } d \in \mathbb{N}$$

Note that in this case (b, d) is even F_B-distinguished. In case we want to explicitly specify the corresponding distinguished basis, we will say that F_B is F_A-*minimal with respect to* (b, d). We will see below that tangential flatness of $f : (A, F_A) \to (B, F_B)$ implies that F_B is F_A-minimal.

5.5. Distinguished bases and flatness

Let $f : (A, m) \to (B, n)$ be a local homomorphism of local rings, $\bar{F}$ a filtration on the special fibre $\bar{B} = B/mB$ of f, and (b, d) an $\bar{F}$-distinguished basis of B over A. Then

(i) The homomorphism

(1) $$\phi : \prod_{d \in \mathbb{N}} (\bigoplus_{d(\lambda) = d} A)^\wedge \to B^\wedge, \quad (a_\lambda)_{\lambda \in \Lambda} \mapsto \sum_{\lambda \in \Lambda} a_\lambda \, b(\lambda)$$

is well-defined and surjective.

(ii) The homomorphism ϕ of (i) is bijective if and only if the homomorphism $f : (A, m) \to (B, n)$ is flat.

Here "$\wedge$" denotes m-adic completion on the left hand side of formula (1) and n-adic completion on the right.

Proof. Step 1: Reduction to the complete case. Let $f^\wedge : A^\wedge \to B^\wedge$ be the homomorphism induced by f on the completions with respect to the natural topologies. Recall that in 4.1 we describe the completion of a free A-module as a certain submodule of some direct product. Using this description, we see that the homomorphism ϕ of (i) is the same for f and $f^\wedge$. Moreover, f is flat if and only if so is $f^\wedge$ (cf [Ma86], Th. 22.4(i)). Passage to the completions doesn't have any effect on the graded ring associated with $\bar{B}$, i.e., (b, d) is also a distinguished basis for $f^\wedge$. This reduces the proof to the complete case.

Step 2: The homomorphism ϕ of (i) is well-defined.

To prove that the homomorphism

$$\phi : \prod_{d \in \mathbb{N}} (\bigoplus_{\mathrm{d}(\lambda)=d} A)^\wedge \to B, \quad (a_\lambda)_{\lambda \in \Lambda} \mapsto \sum_{\lambda \in \Lambda} a_\lambda \, \mathrm{b}(\lambda)$$

is well-defined, identify the m-adic completion

$$(\bigoplus_{\mathrm{d}(\lambda)=d} A)^\wedge = \lim_d \mathrm{proj}(\bigoplus_{\mathrm{d}(\lambda)=d} A) \otimes_A A/m^{d+1}$$

with a submodule of the direct product $\prod_{\mathrm{d}(\lambda)=d} A$ (see 4.1). Then the elements of the module $\prod_{d \in \mathbb{N}} (\bigoplus_{\mathrm{d}(\lambda)=d} A)^\wedge$ are certain families $(a_\lambda)_{\lambda \in \Lambda} \in \prod_{\lambda \in \Lambda} A$. We have to show that the series $\sum_{\lambda \in \Lambda} a_\lambda \, \mathrm{b}(\lambda)$ are n-adically convergent for these families. To do so, it is sufficient to show that for every $d \in \mathbb{N}$ there is only a finite number of $\lambda \in \Lambda$ with $a_\lambda \, \mathrm{b}(\lambda)$ not in n^d. From the definition of a distinguished basis we see that there is some $k \in \mathbb{N}$ such that $\mathrm{b}(\lambda) \in n^d$ for $\mathrm{d}(\lambda) \geq k$. So it is sufficient to consider those $\lambda \in \Lambda$ satisfying $\mathrm{d}(\lambda) < k$. The description of the modules $(\bigoplus_{\mathrm{d}(\lambda)=d} A)^\wedge$ as submodules of a direct product (see 4.1) shows that

$$a_\lambda \in m^d \quad (\text{hence } a_\lambda \, \mathrm{b}(\lambda) \in n^d)$$

for all but a finite number of λ with $\mathrm{d}(\lambda) < k$. Thus the homomorphism ϕ is well-defined.

Step 3: Surjectivity of the homomorphism ϕ of (i). We will first prove that the homomorphism

$$\bar{\phi} : \prod_{d \in \mathbb{N}} (\bigoplus_{\mathrm{d}(\lambda)=d} K) \to \bar{B}, \quad (c_\lambda)_{\lambda \in \Lambda} \mapsto \sum_{\lambda \in \Lambda} c_\lambda \bar{\mathrm{b}}(\lambda)$$

with $K := A/m$ is bijective. One might consider this as a special case of the theorem (when A is a field). From the definition of b : $\Lambda \to B$ we see that

$$\bar{F}^d = \sum_{\mathrm{d}(\lambda)=d} K \cdot \bar{\mathrm{b}}(\lambda) + \bar{F}^{d+1}$$

and

(2) $\qquad \sum_{\lambda \in \Lambda} c_\lambda \bar{\mathrm{b}}(\lambda) \in \bar{F}^d$ with $c_\lambda \in K$ implies $c_\lambda = 0$ for $\mathrm{d}(\lambda) < d$.

The first relation means that, given any $y \in \bar{B}$, there is a sequence of elements

$$x_d = \sum_{d(\lambda)=d} c_\lambda \bar{b}(\lambda) \in \sum_{d(\lambda)=d} K \cdot \bar{b}(\lambda) \quad (d \in \mathbb{N})$$

such that $y - (x_0 + x_1 + \ldots + x_d) \in F_B^{d+1} \cdot \bar{B}$ for every $d = 0,1,2,\ldots$ Thus the series $\sum x_d$ converges to y. By construction, $\bar{\phi}((c_\lambda)_{\lambda \in \Lambda}) = \sum c_\lambda b(\lambda) = \sum x_d = y$. This proves that $\bar{\phi}$ is surjective. From (2) above we see that the map is even bijective.

To prove surjectivity of ϕ, we will apply the filtered Grothendieck lemma 2.5 with $I = m$. For this purpose the modules in question must be equipped with filtrations. In that what follows, the A-module

$$X := \prod_{d \in \mathbb{N}} (\bigoplus_{d(\lambda)=d} A)^{\wedge}$$

will be equipped with the filtration

$$F_X^d := X \cap \left(\prod_{n \in \mathbb{N}} \prod_{d(\lambda)=n} m^{d-n} \right).$$

Note that

$$\lim_d \mathrm{proj}\, X/F_X^d = \lim_d \mathrm{proj} \prod_{n \in \mathbb{N}} (\bigoplus_{d(\lambda)=n} A/m^{d-n})$$

$$= \prod_{n \in \mathbb{N}} \lim_d \mathrm{proj}(\bigoplus_{d(\lambda)=n} A/m^{d-n})$$

$$= X.$$

So (X, F_X) is complete. Next equip B with the n-adic topology. Then B is also m-adically complete (see 2.9) and the homomorphism ϕ is a homomorphism of complete filtered modules over A. To prove that it is surjective, it is sufficient to show that $\phi \otimes_A A/m$ is surjective (by the filtered Grothendieck Lemma 2.5). We will even prove that $\phi \otimes_A A/m$ is an isomorphism, since this will be needed for the proof of assertion (ii). Consider the commutative diagram

$$
\begin{array}{ccc}
X & \xrightarrow{\phi} & B \\
\downarrow & & \downarrow \\
\prod_{d \in \mathbb{N}} (\oplus_{d(\lambda)=d} K) & \xrightarrow{\bar{\phi}} & B/mB
\end{array}
$$

with the vertical surjections being the natural ones. Tensor the upper row with A/m to get a commutative diagram

$$
\begin{array}{ccc}
X/mX & \xrightarrow{\phi \otimes_A A/m} & B/mB \\
\alpha \downarrow & & \| \\
\prod_{d \in \mathbb{N}} (\oplus_{d(\lambda)=d} K) & \xrightarrow{\bar{\phi}} & B/mB
\end{array}
$$

We already know that $\bar{\phi}$ is bijective. So $\phi \otimes_A A/m$ is surjective, and it will be sufficient to show that the left vertical surjection α is injective. Since the ideal m is finitely generated, the functor $? \otimes_A A/m$ commutes with direct products, i.e., it is sufficient to show that the natural homomorphisms

$$(\bigoplus_{d(\lambda)=d} A)^\wedge \otimes_A A/m \rightarrow (\bigoplus_{d(\lambda)=d} A/m)$$

are injective. In other words we have to prove the injectivity of the canonical map

$$Y^\wedge/mY^\wedge \rightarrow Y/mY$$

for free A-modules Y that are equipped with the filtration $F_Y^d := F_A^d \cdot Y$. But this is just the assertion of 1.18, for F_Y induces on Y/mY the discrete topology, i.e., $Y/mY = (Y/mY)^\wedge$. Note that F_Y is Artin-Rees by 1.15(iv).

Step 4: The homomorphism ϕ of (i) is bijective in case f is flat. By the filtered Grothendieck Lemma 2.5 this follows from what we already have proved.

Step 5: f is flat if the homomorphism ϕ of (i) is bijective.

This follows from the facts that the completion of a free module is flat (see 4.1(ii)) and that a direct product of flat modules is flat ([Ma86], Ex 7.4).

5.6. Convention: the completion of a flat extension as a direct product

Let $f : (A, m) \rightarrow (B, n)$, $\bar{F}$ and (b, d) be as in the theorem above, i.e., $f : (A, m) \rightarrow (B, n)$ is a flat local homomorphism of local rings, $\bar{F}$ a filtration on the special fibre $\bar{B} = B/mB$ of f and (b, d) an F-distinguished basis of B over A. Then we will use the homomorphism ϕ of the theorem above to identify the completion of B with the direct product

$$\prod_{d \in \mathbb{N}} (\bigoplus_{d(\lambda)=d} A)^\wedge$$

("$\wedge$" denotes m-adic completion). To stress this identification, the direct product will be sometimes written

$$\overset{\wedge}{\underset{\lambda \in \Lambda}{\sum}} A^\wedge \cdot b(\lambda)$$

i.e., we use a hat on the sum sign to indicate that the expression means a module consisting of (convergent) infinite sums. If $(I_\lambda)_{\lambda \in \Lambda}$ is a family of ideals $I_\lambda \subseteq A^\wedge$, we will write

$$\overset{\wedge}{\underset{\lambda \in \Lambda}{\sum}} I_\lambda \cdot b(\lambda) := B^\wedge \cap \prod_{n \in \mathbb{N}} (\prod_{d(\lambda)=n} I_\lambda)$$

for the $A^\wedge$-module of convergent series $\sum_{\lambda \in \Lambda} a_\lambda b(\lambda)$ such that the coefficient a_λ is in I_λ for every λ. This is just the closure of the usual sum $\sum_{\lambda \in \Lambda} I_\lambda \cdot b(\lambda)$ in

$B^{\wedge} = \prod_{d \in \mathbb{N}} (\oplus_{d(\lambda)=d} A)^{\wedge}$ with respect to the topology defined by the filtration
$F^d := B^{\wedge} \cap \prod_{n \in \mathbb{N}} (\prod_{d(\lambda)=n} m^{d-n})$, for, letting $N := \sum_{\lambda \in \Lambda} I_\lambda \cdot b(\lambda)$ and $M :=$

$\underset{d \in \mathbb{N}}{\oplus}$ $(\oplus_{d(\lambda)=d} A^{\wedge})$ the description of the closure $\mathrm{cl}(N)$ in 1.16 yields,

$$\mathrm{cl}(N) = \mathrm{Ker}(M^{\wedge} \to (M/N)^{\wedge}),$$

where

$$(M/N)^{\wedge} = (\underset{d \in \mathbb{N}}{\oplus} (\underset{d(\lambda)=d}{\oplus} A^{\wedge}/I_\lambda))^{\wedge} = \prod_{d \in \mathbb{N}} (\underset{d(\lambda)=d}{\oplus} A^{\wedge}/I_\lambda)^{\wedge}.$$

Therefore, $\mathrm{cl}(N) = \prod_{d \in \mathbb{N}} (\oplus_{d(\lambda)=d} A)^{\wedge} \cap \prod_{n \in \mathbb{N}} (\prod_{d(\lambda)=n} I_\lambda)$, as stated above.

5.7. Structure constants

Let $f : (A, m) \to (B, n)$ be a local homomorphism of complete local rings, $\bar{F}$ a filtration on the special fibre $\bar{B} = B/mB$ of f, and (b, d) an $\bar{F}$-distinguished basis of B over A. Then, by the above theorem, there are elements $a_{\alpha\beta}^{\gamma} \in A$ such that

$$b(\alpha) \cdot b(\beta) = \sum_{\gamma \in \Lambda} a_{\alpha\beta}^{\gamma} \, b(\gamma) \text{ for arbitrary } \alpha, \beta \in \Lambda.$$

These elements are called *structure constants* of B over A with respect to (b, d). Note that the structure constants are uniquely determined by (b, d) in case that B is flat over A.

Remark

The inequalities

$$\mathrm{ord}(a_{\alpha\beta}^{\gamma}) \geq d(\alpha) + d(\beta) - d(\gamma)$$

will play an important role below. Note that, if these inequalities are satisfied, the F_A-minimal ring filtration

$$F_B^{'d} := \sum_{(i,\lambda) \in \mathbb{N}^n \times \Lambda^n} F_A^{d - \langle i, d(\lambda) \rangle} \cdot B \cdot b(\lambda)^i$$

is equal to the module filtration

$$F_B^{''d} := \sum_{\lambda \in \Lambda} F_A^{d - d(\lambda)} \cdot B \cdot b(\lambda).$$

Both filtrations are minimal with respect to the properties that they contain $F_A \cdot B$ and that $b(\lambda)$ has order at least $d(\lambda)$. Their equality is just a consequence of the fact that the conditions on the order of the structure constants imply that F_B'' is a ring filtration:

$$F_B''^d \cdot F_B''^{d'} \subseteq \sum_{\lambda,\mu \in \Lambda} F_A^{d-d(\lambda)} F_A^{d'-d(\mu)} \cdot B \cdot b(\lambda) b(\mu)$$

$$\subseteq \sum_{\lambda,\mu,\alpha \in \Lambda} F_A^{d-d(\lambda)} F_A^{d'-d(\mu)} a_{\lambda\mu}^{\alpha} \cdot B \cdot b(\alpha)$$

$$\subseteq F_B''^{d+d'}.$$

5.8. Distinguished bases and tangential flatness

Let $f : (A, m) \to (B, n)$ be a flat homomorphism of complete filtered local rings, (b, d) a distinguished basis for f, and $(a_{\alpha\beta}^{\gamma})$ the family of structure constants of B over A with respect to (b, d). Then the following are equivalent.

(i) f is tangentially flat.

(ii) $\mathrm{ord}_{F_A}(a_{\alpha\beta}^{\gamma}) \geq d(\alpha) + d(\beta) - d(\gamma)$ for arbitrary $\alpha, \beta, \gamma \in \Lambda$ and

$$F_B^d = \sum_{\lambda \in \Lambda}^{\wedge} F_A^{d-d(\lambda)} \cdot b(\lambda)$$

for arbitrary $d \in \mathbb{N}$.

(iii) $\mathrm{ord}_{F_A}(a_{\alpha\beta}^{\gamma}) \geq d(\alpha) + d(\beta) - d(\gamma)$ for arbitrary $\alpha, \beta, \gamma \in \Lambda$ and F_B is F_A-minimal with respect to (b, d).

Proof. Step 1: The inequalities $\mathrm{ord}_{F_A}(a_{\alpha\beta}^{\gamma}) \geq d(\alpha) + d(\beta) - d(\gamma)$ imply that the A-modules

$$F(d) := \sum_{\lambda \in \Lambda}^{\wedge} F_A^{d-d(\lambda)} b(\lambda)$$

satisfy $F(d) \cdot F(d') \subseteq F(d + d')$ and $F(d) = \sum_{\lambda \in \Lambda} F_A^{d-d(\lambda)} B \cdot b(\lambda)$ for arbitrary d and d'.

Take elements $b \in F(d)$ and $b' \in F(d')$ and write them as convergent series.

$$b = \sum_{\lambda \in \Lambda} a_\lambda b(\lambda), \qquad a_\lambda \in F_A^{d-d(\lambda)}$$

$$b' = \sum_{\lambda \in \Lambda} a_\lambda' b(\lambda), \qquad a_\lambda' \in F_A^{d'-d(\lambda)}$$

67

Their product is equal to the following convergent series (calculate modulo n^k and let then k go to infinity):

$$b \cdot b' = \sum_{\alpha,\beta \in \Lambda} a_\alpha \, b(\alpha) a'_\beta \, b(\beta) = \sum_{\alpha,\beta,\lambda \in \Lambda} a_\alpha a'_\beta a^\lambda_{\alpha\beta} \, b(\lambda)$$

By assumption,

$$a_\alpha a'_\beta a^\lambda_{\alpha\beta} \in F_A^{d-d(\alpha)} \cdot F_A^{d'-d(\beta)} \cdot F_A^{d(\alpha)+d(\beta)-d(\gamma)} \subseteq F_A^{d+d'-d(\gamma)},$$

hence $bb' \in F(d+d')$. This proves $F(d) \cdot F(d') \subseteq F(d+d')$. By 5.5(i),

$$F(0) = B = F_B^0.$$

Therefore $B \cdot F(d) = F(0) \cdot F(d) \subseteq F(d)$, i.e., the A-modules $F(d)$ are even ideals of B. In particular, $\sum_{\lambda \in \Lambda} F_A^{d-d(\lambda)} B \cdot b(\lambda) \subseteq F(d)$. The converse inclusion follows from the fact that $\sum_{\lambda \in \Lambda} F_A^{d-d(\lambda)} B \cdot b(\lambda)$ is a closed subset of B.

Step 2: Proof of (i)$\Rightarrow$(ii). By 5.3, the family $(\text{in}(b(\lambda)))_{\lambda \in \Lambda}$ is a free system of generators of $G(B)$ over $G(A)$. From the defining identities,

$$\sum_{\lambda \in \Lambda} a^\gamma_{\alpha\beta} \, b(\lambda) = b(\alpha) \cdot b(\beta) \quad (\in F_B^{d(\alpha)} \cdot F_B^{d(\beta)} \subseteq F_B^{d(\alpha)+d(\beta)})$$

we deduce that the structure constants satisfy $a^\gamma_{\alpha\beta} \in F_B^{d(\alpha)+d(\beta)-d(\gamma)}$ (cf 2.6). Note that, since F_B is cofinite, there are at most finitely many terms in the sum on the left which are not in $F_B^{d(\alpha)+d(\beta)}$ (by 5.1 condition 3 and 4.1(i), cf step 2 in the proof of 5.5). We have proved, the inequalities

$$\text{ord}_{F_A}(a^\gamma_{\alpha\beta}) \geq d(\alpha) + d(\beta) - d(\gamma)$$

are satisfied, and hence, by step 1, the A-submodules $F(d) \subseteq B$ are even ideals of B, hence are closed in the topology defined by F_B. Moreover, since F_B^d is a closed set and since (b,d) is F_B-distinguished, i.e., $d(\lambda) = \text{ord}_{F_B} b(\lambda)$,

$$F(d) = \sum_{\lambda \in \Lambda}^{\wedge} F_A^{d-d(\lambda)} \, b(\lambda) \subseteq F_B^d.$$

The filtrations, F_B and $(F(d))_{d \in \mathbb{N}}$ induce the same filtration on the special fibre $\bar{B} = B/mB$ of f. For, since (b,d) is an F_B-distinguished basis,

$$F_B^d \cdot \bar{B} = \sum_{d(\lambda) \geq d} (A/m) \cdot \bar{b}(\lambda) + F_B^{d+1} \cdot \bar{B}$$

i.e., every element of $F_B^d \cdot \bar{B}$ can be written as a convergent series from

$$\sum_{d(\lambda) \geq d}^{\wedge} (A/m) \cdot \bar{b}(\lambda),$$

hence is the residue class of an element from $F(d)$. Thus flatness of $G(B)$ over $G(A)$ implies (by 3.12)

$$F_B^d = F(d) = \sum_{\lambda \in \Lambda}^{\wedge} F_A^{\prime d - d(\lambda)} b(\lambda).$$

Step 3: Proof of (ii)$\Rightarrow$(iii). This implication follows directly from the first step (cf the remark of 5.7).

Step 4: Proof of (iii)$\Rightarrow$(i). By the first step and the remark of 5.7,

$$F(d) = \sum_{\lambda \in \Lambda} F_A^{d - d(\lambda)} B \cdot b(\lambda) = F_B^d \text{ for every } d.$$

So the graded ring $G(B)$ can be written as follows.

$$G(B) = \bigoplus_{d=0}^{\infty} F(d)/F(d+1) = \bigoplus_{d=0}^{\infty} \prod_{n \in \mathbb{N}} (\bigoplus_{d(\lambda)=n} F_A^{d-n}/F_A^{d+1-n})$$

$$= \bigoplus_{d=0}^{\infty} \bigoplus_{\lambda \in \Lambda} F_A^{d-d(\lambda)}/F_A^{d+1-d(\lambda)} = \bigoplus_{\lambda \in \Lambda} G(A)[-d(\lambda)]$$

Note that in the above calculation the direct product can be replaced by a direct sum, since, for every fixed d, there are only finitely many non-zero direct factors under the product sign. We have proved that B is tangentially flat over A (even 'tangentially free')

5.9. Cohen factorization

Let $f : (A, m) \to (B, n)$ be a homomorphism of complete filtered local rings such that the filtration F_B of B is q-adic over F_A for some N-tuple of positive real numbers $q = (q_1, \ldots, q_N)$, i.e., there is an N-tuple $b = (b_1, \ldots, b_N)$ of elements from n such that

$$F_B^d := \sum_{i + <q, j> \geq d} F_A^i b^j B \quad \text{for every } d \in \mathbb{N}.$$

Then there is a commutative diagram of homomorphisms of complete filtered local rings

$$\begin{array}{ccc} (A, m) & \xrightarrow{f} & (B, n) \\ g \downarrow & \nearrow h & \\ (R, p) & & \end{array}$$

such that

(i) g is tangentially flat.

(ii) h is surjective.

(iii) The filtration of R is q'-adic over F_A with direct image filtration $h_* F_R = F_B$. Here, as usual, we are using multi index notation, i.e.,

$$< q, j > := q_1 j_1 + \ldots + q_N j_N \text{ and } b^j := b_1^{j_1} \cdot \ldots \cdot b_N^{j_N}$$

if $j = (j_1, \ldots, j_N)$. Note that we do not claim that the weight vector q' for the filtration of R is equal to q.

Proof. Step 1: Reduction to the case that $b_1, \ldots, b_N$ generate n.

Given any non-zero element $\beta \in n$, let r denote the F_B-order of β so that

$$\beta \in F_B^r.$$

Define $b^* := (\beta, b_1, \ldots, b_N)$, $q^* := (r, q_1, \ldots, q_N)$, and

$$F^*{}_B^d := \sum_{i + <q^*, j^*> \geq d} F_A^i b^{*j^*} B.$$

Then, $F_B \subseteq F_B^*$, since the power products b^j in the definition of F_B^d are exactly the power products b^{*j^*} in the definition of $F^*{}_B^d$ such that the first coordinate of j^* is zero. On the other hand, if $j^* := (s, j_1, \ldots, j_N)$ satisfies the condition

$$i + <q^*, j^*> \geq d,$$

then, letting $j = (j_1, \ldots, j_N)$,

$$F_A^i b^{*j^*} B = F_A^i \beta^s b^j B \subseteq F_B^{i+sr+[<q,j>]} = F_B^{i+[<q^*,j^*>]} \subseteq F_B^d.$$

Here $[x]$ denotes the largest integer $\leq x$. We have proved $F_B^* = F_B$. In other words, we can add an arbitrary element $\beta \in n$ to the coordinates of b without changing the filtration F_B. Therefore, we may assume that the coordinates of b generate the maximal ideal n of B.

Step 2: The proof in case the b_i's generate n.

By Cohen structure theory ([G-D64], EGA IV_1, §19) there are Cohen rings C_A, C_B (see [G-D64], (19.8.4)) and local homomorphisms $C_A \to A$ and $C_B \to B$ inducing isomorphisms of the residue class fields (see [G-D64], (19.8.6)(ii) and (i)). Adjoining sets of indeterminates $X = \{X_1, .., X_r\}$ and $Y = \{Y_1, \ldots, Y_s\}$, one obtains surjections

$$C_A[[X]] \to A \text{ and } C_B[[Y]] \to B.$$

Note that is is sufficient to choose indeterminates such that they map onto a generating system of the respective maximal ideals. In particular, we may assume that the number of indeterminates in the set Y is N and that Y_i maps to $b_i \in B$ for every i. The two maps fit into a commutative diagram

$$
\begin{array}{ccc}
A & \stackrel{f}{\longrightarrow} & B \\
\uparrow & & \uparrow \\
C_A[[X]] & & C_B[[Y]] \\
\cup & & \cup \\
C_A & & C_B \\
\uparrow & \nearrow & \\
\mathbb{Z}_{(\pi)} & &
\end{array}
$$

where $\mathbb{Z}_{(\pi)}$ is the localization of the integers $\mathbb{Z}$ at the characteristics π of A/m in case this characteristics is positive. In the characteristics zero case let $\mathbb{Z}_{(\pi)}$ denote the rational numbers. The maps from $\mathbb{Z}_{(\pi)}$ to C_A and C_B are the uniquely determined local homomorphisms preserving the identity element. Since C_A, as a Cohen ring, is formally smooth over $\mathbb{Z}_{(\pi)}$ (see [G-D64], (19.7.1) and (19.6.1)), there is a homomorphism $C_A \to C_B[[Y]]$ that can be commutatively included into this diagram, hence there is a commutative diagram of local homomorphisms

$$
\begin{array}{ccc}
A & \stackrel{f}{\longrightarrow} & B \\
p_A \uparrow & & \uparrow p_B \\
C_A[[X]] & \stackrel{\varphi}{\longrightarrow} & C_B[[X,Y]]
\end{array}
$$

where φ induces the identity map on the set X. Define

$$
R := C_B[[X,Y]]/(\varphi(\operatorname{Ker} p_A))
$$

and use p_A to identify A with $C_A[[X]]/\operatorname{Ker}(p_A)$. Tensoring the lower row of the diagram with A over $C_A[[X]]$, one obtains a commutative diagram

$$
\begin{array}{ccc}
A & \stackrel{f}{\longrightarrow} & B \\
& g \searrow & \uparrow h \\
& & R
\end{array}
$$

where $g := \varphi \otimes_{C_A[[X]]} A$ and h is induced by p_B. Equip R with the filtration F_R such that

$$
F_R^d := \sum_{i+<q,j>\geq d} F_A^i \cdot Y^j \cdot R.
$$

Then g and h are homomorphisms of filtered local rings, and the filtration of B is the direct image filtration of F_R. By construction, h is surjective. So all we have yet to show is that g is tangentially flat. As above, let π denote the characteristics of A/m, such that the natural image of π in C_A and C_B generates the maximal ideal of these rings. Then $\text{Ker}(p_A) \subseteq (\pi, X)C_A[[X]]$ hence

$$R/mR \cong C_B[[X, Y]]/(\pi, X) \cong L[[Y]],$$

where L denotes the residue class field of B. Choose a family $(\beta_\gamma)_{\gamma \in \Gamma}$ of units from R representing a vector space basis of L over the residue class field $K :=$ A/m. Then the maps $\mathrm{b} : \Lambda \to R$ and $\mathrm{d} : \Lambda \to \mathbb{N}$ with

$$\Lambda := \Gamma \times \mathbb{N}^N, \quad \mathrm{b}(\gamma, j) := \beta_\gamma Y^j \bmod(\varphi(\text{Ker } p_A)), \quad \mathrm{d}(\gamma, j) := [< q, j >]$$

form an F_R-distinguished basis of R over A. To see this, we have to show that this is an F_S-distinguished basis where F_S denotes the filtration induced by F_R on $S := L[[Y]]$. Note that

$$F_S^d = \sum_{<q,j> \geq d} Y^j S.$$

hence $G_{F_S}(S)(d) \cong \bigoplus_{[<q,j>]=d} Y^j L$. In particular, $\beta_\gamma Y^j$ has F_S-order $[< q, j >]$ and the initial forms of these elements form a vector space basis of $G_{F_S}(S)$ over K. So the pair (b, d) is indeed F_S-distinguished. The third defining condition of 5.1 on a distinguished basis is satisfied, since the coordinates of q are are bounded, say $q_i \leq \rho$ for every i. For, letting $k(d) := \rho \cdot N \cdot d$, the condition $\mathrm{d}(\gamma, j) \geq k(d)$ implies

$$N \cdot \rho \cdot \max(j_i) \geq [< q, j >] = \mathrm{d}(\gamma, j) \geq k(d) = \rho \cdot N \cdot d,$$

hence $j_i \geq d$ for at least one i. Therefore $\mathrm{b}(\gamma, j) = \beta_\gamma \cdot Y^j$ is in the d-th power of the maximal ideal, as required.

Since the order of an element cannot decrease under local homomorphisms,

$$\text{ord}_{F_R}(\mathrm{b}(\gamma, j)) \geq [< q, j >] = \text{ord}_{F_S}(\beta_\gamma X^j).$$

So the basis is also F_R-distinguished. To prove that $g : A \to R$ is tangentially flat, it will be sufficient to show that the structure constants belonging to (b, d) satisfy the conditions of 5.8. Note that g is flat, for, $\varphi : C_A[[X]] \to C_B[[X, Y]]$ is flat ([Ma86], Ex. 22.3), since it maps the regular system of of parameters $(\pi,)X_1, \ldots, X_r$ into a regular sequence, and g is just $\varphi \otimes A$. By construction, F_R is F_A-minimal. To check the condition on the orders of the structure constants, write in R,

$$\beta_{\gamma'} \cdot \beta_{\gamma''} = \sum_{\gamma \in \Gamma, j \in \mathbb{N}^N} \alpha_{\gamma', \gamma''}^{\gamma, j} \beta_\gamma y^j \text{ with } \alpha_{\gamma', \gamma''}^{\gamma, j} \in A.$$

where y^j denotes the residue class in R of $Y^j \in C_B[[X, Y]]$. Then,

$$\beta_{\gamma'} y^{j'} \cdot \beta_{\gamma''} y^{j''} = \sum_{\gamma \in \Gamma, j \in \mathbb{N}^N} \alpha_{\gamma', \gamma''}^{\gamma, j} \beta_\gamma y^{j + j' + j''} \text{ with } \alpha_{\gamma', \gamma''}^{\gamma, j} \in A.$$

We have to show that the non-zero structure constants satisfy

$$\operatorname{ord}_{F_A} \alpha^{\gamma,j}_{\gamma',\gamma''} \geq d(\gamma',j') + d(\gamma'',j'') - d(\gamma,j+j'+j').$$

But this is trivial, since the expression on the right is ≤ 0,

$$d(\gamma',j')+d(\gamma'',j'')-d(\gamma,j+j'+j') = [<q,j'>]+[<q,j''>]-[<q,j+j'+j''>]$$
$$\leq [<q,j'>+<q,j''>]-[<q,j+j'+j''>] \leq 0.$$

5.10. Tangential flatness and the order function

Let $f : (A,m) \to (B,n)$ be a flat local homomorphism of filtered local rings. Then the following assertions are equivalent.

(i) B is tangentially flat over A.

(ii) The following two conditions are satisfied.

    a) $\operatorname{ord}_B(ab) \leq \operatorname{ord}_A(a) + \operatorname{ord}_B(b)$ for every two elements $a \in A$ and $b \in B$ such that

$$\operatorname{ord}_B(b) = \operatorname{ord}_{B/mB}(b \bmod mB)$$

    (in which case equality holds).

    b) Property (a) is preserved under the substitution

$$f \mapsto f \otimes_A A/I$$

for every proper ideal I of A.

Proof. (i)$\Rightarrow$(ii). Assume there are elements $a \in A$ and $b \in B$ such that

$$\operatorname{ord}(ab) > \operatorname{ord}(a) + \operatorname{ord}(b) \quad \text{and} \quad \operatorname{ord}(b) = \operatorname{ord}(b \bmod mB).$$

Replacing A and B by A/am and B/amB, respectively, we may assume

$$a \cdot m = 0.$$

Note that such a substitution cannot decrease $\operatorname{ord}(ab)$ and has no effect on the other two orders in question, $\operatorname{ord}(a)$ and $\operatorname{ord}(b)$. The latter is obvious for $\operatorname{ord}(b)$, since we assume $\operatorname{ord}(b) = \operatorname{ord}(b \bmod mB)$. To see this for $\operatorname{ord}(a)$, suppose

$$a \in F_A^{d+1} + am \quad \text{with} \quad d := \operatorname{ord}(a)(< \infty).$$

Then there is an element $y \in m$ with $(1-y)a \in F_A^{d+1}$. Since A is local, $1-y$ is a unit, i.e., $a \in F_A^{d+1}$, which contradicts the definition of d. Thus we may assume $a \cdot m = 0$ and, in particular,

$$\operatorname{in}(a) \cdot \operatorname{in}(m) = 0.$$

Since $in(m)$ is maximal in $G(A)$, it must be the full annihilator of $in(a)$, i.e., multiplication by $in(a)$ defines an exact sequence

$$0 \to in(m) \to G(A) \xrightarrow{in(a)} G(A).$$

Tensor with $G(B)$ and consider the resulting exact sequence

$$0 \to in(m)G(B) \to G(B) \xrightarrow{in(a)} G(B).$$

By assumption, $\mathrm{ord}(ab) > \mathrm{ord}(a) + \mathrm{ord}(b)$, hence $in(a) \cdot in(b) = 0$. The element $in(b)$ is in the annihilator of $in(a)$ in $G(B)$, i.e., in the kernel of the canonical mapping

$$G(B) \to G(B)/in(m)G(B) \cong G(B/mB).$$

But then $\mathrm{ord}(b) < \mathrm{ord}(b \bmod mB)$ contrary to our assumption. We have proved, (i) implies condition (a) of (ii). Condition (ii)(b) follows from the fact that tangential flatness is preserved under surjective base change (by 3.8(i)).

(ii)$\Rightarrow$(i). Since tangential flatness is an infinitesimal property (by 3.16), we may assume that (A, m) is Artinian ring. Let S be the set of proper ideals J of A such that $G(B/JB)$ is not flat over $G(A/J)$. We want to show S is empty. Suppose it is not. Since A is Noetherian there is some $J \in S$ being maximal with respect to inclusion. Replacing A and B by A/J and B/JB, respectively, we may assume the only element in S is the zero ideal,

$$S = \{ (0) \}.$$

Since $G(B)$ is not flat over $G(A)$, there is a proper ideal J of A such that

$$in(JB) \neq in(J) \cdot G(B)$$

(see 4.3). Obviously $J \neq 0$. Let t be some non-zero element from J,

$$0 \neq t \in J.$$

Since tA is not in S, $G(B/tB)$ is $G(A/tA)$-flat, hence

$$in(J \cdot (B/tB)) = in(J/tA) \cdot G(B/tB),$$

i.e.,

$$in(JB) = in(J) \cdot G(B) + in(tB).$$

This shows, $in(tB) \neq in(tA) \cdot G(B)$ for every non-zero element $t \in J$. Since A is supposed to be Artinian, it is possible to choose $t \in J$ such that $L(tA) = 1$, i.e., $t \cdot m = 0$ hence $t \cdot mB = 0$.

As we have seen there is some $\alpha \in in(tB) - in(tA)G(B)$. We may assume that α is homogeneous, say,

$$\alpha = in(tb) \quad \text{with} \quad b \in B.$$

Since mB annihilates t, the product tb is independent upon the special choice of b inside its residue class modulo mB. So we may assume

$$\mathrm{ord}(b) = \mathrm{ord}(b \bmod mB).$$

But then, by assumption, $\mathrm{ord}(tb) = \mathrm{ord}(t) + \mathrm{ord}(b)$, hence

$$\alpha = (tb \bmod F_B^{\mathrm{ord}(tb)+1}) = (t \bmod F_A^{\mathrm{ord}(t)+1}) \cdot (b \bmod F_B^{\mathrm{ord}(b)+1})$$
$$= \mathrm{in}(t) \cdot \mathrm{in}(b)$$
$$\in \mathrm{in}(tA) \cdot G(B)$$

contradicting the choice of α. This proves that $G(B)$ is $G(A)$-flat.

5.11. Tangential flatness as an infinitesimal Property II

Let $f : (A, m) \to (B, n)$ be a homomorphism of filtered local rings and $(I_\lambda)_{\lambda \in \Lambda}$ a directed family of ideals in A such that

1. Each F_A^d is closed in the topology defined by the I_λ's (see [Ma86], §8).

2. $\quad \bigcap_{\lambda \in \Lambda} I_\lambda = 0.$

Then the following conditions are equivalent.

(i) f is tangentially flat.

(ii) $f \otimes_A A/I_\lambda : A/I_\lambda \to B/I_\lambda B$ is tangentially flat for every λ.

Proof. Implication (i)$\Rightarrow$(ii) follows from 3.8. So assume $f \otimes_A A/I_\lambda$ is tangentially flat for every λ. Since by 3.8 our assumption (ii) is preserved under the substitution $f \mapsto f \otimes_A A/I$, it will be sufficient to show that condition (ii)(a) of 5.10 is satisfied. Fix elements $a \in A$ and $b \in B$ with

$$\mathrm{ord}_B(b) = \mathrm{ord}_{B/mB}(b \bmod mB).$$

We have to show

$$\mathrm{ord}_B(ab) \leq \mathrm{ord}_A(a) + \mathrm{ord}_B(b).$$

In case a or b is zero, this is trivial, so assume $a \neq 0$ and $b \neq 0$. In particular, a has finite F_A-order, say d. Since F_A^{d+1} is closed in the topology defined by the ideals I_λ, there exists some λ with $a \notin F_A^{d+1} + I_\lambda$, i.e.,

(1) $$\mathrm{ord}_{A/I_\lambda}(a) = d = \mathrm{ord}_A(a)$$

where the expression on the left denotes the order of the residue class in A/I_λ of a. Since $f \otimes_A A/I_\lambda$ is tangentially flat,

$$\mathrm{ord}_B(ab) \leq \mathrm{ord}_{B/I_\lambda B}(ab)$$
$$\leq \mathrm{ord}_{A/I_\lambda}(a) + \mathrm{ord}_{B/I_\lambda B}(b)$$
$$= \mathrm{ord}_A(a) + \mathrm{ord}_B(b).$$

Here as above, the function $\mathrm{ord}_{B/I_\lambda B}$ denotes the order of the residue class in $B/I_\lambda B$ of its argument. Note that the inequality in the middle uses 5.10, and the bottom identity comes from (1). The proof is complete.

6. Hilbert series

6.1. Hilbert series of a graded ring

Let G be a graded ring and M a graded G-module such that

(1) $$L_{G(0)}(M(d)) < \infty \text{ for every } d.$$

Then the *Hilbert series* of M is defined to be the formal Laurent series with integer coefficients,

$$\mathrm{H}_M := \mathrm{H}_M^0 := \sum_{d=-\infty}^{\infty} \mathrm{H}_M^0(d) \cdot T^d \in \mathbb{Z}((T))$$

such that

$$\mathrm{H}_M^0(d) := L_{G(0)}(M(d)).$$

Here $\mathbb{Z}((T))$ denotes the ring of formal Laurent series over $\mathbb{Z}$, i.e., the quotient ring of the formal power series ring $\mathbb{Z}[[T]]$ with respect to the powers of the indeterminate T. Note that in the definition of H_M the number of terms with negative d is finite in view of our definition of graded modules in 1.1. Below we shall refer to a graded module satisfying condition (1) as to a graded module with *finite Hilbert function* . The i-th *sum transform* $\mathrm{H}^{(i)}$ of the formal Laurent series $\mathrm{H} \in \mathbb{Z}((T))$ is the series $\mathrm{H}^{(i)} := (1-T)^{-i} \cdot \mathrm{H}$. In case $\mathrm{H} = \mathrm{H}_M$ we simply shall write

$$\mathrm{H}_M^i := (\mathrm{H}_M)^{(i)}.$$

The series H_M^1 is also called *Hilbert-Samuel series*.

6.2. Comparing Laurent series

Let $\mathrm{H}_i := \sum \mathrm{H}_i(d) \cdot T^d$, $i = 1, 2$, be two formal Laurent series with integer coefficients. Then we shall write

$$\mathrm{H}_1 \leq \mathrm{H}_2 \text{ iff } \mathrm{H}_1(d) \leq \mathrm{H}_2(d) \text{ for every } d,$$

i.e., "$\leq$" in the context of Laurent series will always denote the *total order*.

6.3. Hilbert series of a filtered local ring

If (A, m) is a filtered local ring, the *Hilbert series* of A is defined to be the Hilbert series of the associated graded ring $G(A)$ and is denoted

$$\mathrm{H}_A := \mathrm{H}_A^0 := \mathrm{H}_{G(A)}.$$

Similarly we write

$$\mathrm{H}_A^i := (1 - T)^{-i} \cdot \mathrm{H}_A$$

for the *Hilbert-Samuel series* $(i = 1)$ of A and its i-th *sum transform*. Note that our general assumption that ring filtrations are cofinite (see 3.11) implies that the series H_A^i are well-defined. By definition,

$$\mathrm{H}_A^0(d) = L_A(\mathrm{G}(A)(d)) = L_A(F_A^d/F_A^{d+1})$$
$$\mathrm{H}_A^1(d) = L_A(A(d)) = L_A(A/F_A^{d+1})$$
$$\mathrm{H}_A^{i+1}(d) = \sum_{n=0}^{d} \mathrm{H}_A^i(n).$$

6.4. Hilbert series of a homomorphism

Let $f : (A, m) \to (B, n)$ be a homomorphism of filtered local rings. Then

$$\mathrm{H}_f^i := \mathrm{H}_{B/mB}^i$$

is called *Hilbert series* of f in case $i = 0$, *Hilbert-Samuel series* of f in case $i = 1$, and i-th *sum transform* of the Hilbert series of f for arbitrary i. Here the special fibre B/mB of f is as usual equipped with the direct image filtration $F_B \cdot B/mB$.

6.5. The case of natural filtrations

The Hilbert series and its sum transforms of a filtered local ring (A, m) are by far our most important objects. In fact we are mainly interested in the special case when F_A is the natural filtration,

$$F_A^d = m^d \text{ for every } d.$$

In this case H_A is known to involve important information about the local singularity belonging to the ring A. For example, the series H_A can be written as a quotient

$$\mathrm{H}_A := \frac{p(T)}{(1 - T)^n}$$

where the enumerator $p(T)$ is a polynomial with integer coefficients. This presentation of H_A corresponds to the fact that $\mathrm{H}_A(d)$, considered as function of $d \in \mathbb{N}$, is a polynomial function for large d. Lets assume that in the above presentation common factors are canceled. Then n is the pole order of the rational function H_A at $T = 1$ and is equal to the *Krull dimension* $\dim A$ of the singularity. Moreover, the integer $p(1)$ is its *multiplicity*. The latter is easily seen to be always positive and equals up to the factor $n!$ just the highest coefficient of the "polynomial" $H_A(d)$ (see [Ma86] for details). So the asymptotic behavior of the function $d \mapsto H_A(d)$ is understood at least to some extent. Quite the contrary, the behavior of $H_A(d)$ for small values of d is rather mysterious from the point of view of our current knowledge. Many papers have been devoted to the study of the first values of $H_A(d)$ in special situations. The results seem to say, everything is possible for sufficiently general singularities, though one knows from a result of Macaulay that a rational function of the above type has to satisfy strong growth conditions in order to be a Hilbert function. *Lech's problem* is one of the very few general assertions fitting into this context which couldn't be disproved at once. Meanwhile it has survived for more that 30 years. Given a flat local homomorphism $f : (A, m) \rightarrow (B, n)$ of (naturally filtered) local rings the problem is whether there exists a non-negative integer i such that

$$H_A^i(d) \le H_B^i(d) \quad \text{for} \quad d \in \mathbb{N}.$$

The relation is stated for all values of the Hilbert functions, and one can easily see that it is sufficient to give an answer in the case of Artinian local rings A (i.e., is important to know what happens for small d).

The philosophy behind the problem might be formulated as follows. The resolution of singularities tells us that bad singularities have big Hilbert function, the smallest Hilbert function for a given dimension is the Hilbert function of a regular local ring. So Lech's question could be reformulated as whether the singularities of the parameter space $Spec(A)$ of a deformation are never worse than the corresponding singularities of the deformation space $Spec(B)$. This is something which fits very well into our experience. The parameter spaces of deformations studied nowadays have usually rather weak singularities or are non-singular at all. So it is natural to ask whether there is a theorem behind this phenomenon or whether this simply reflects our very restricted knowledge about general singularities.

6.6. Power series extensions

Let (A, m) be a filtered local ring, $X = \{X_1, \ldots, X_N\}$ a finite set of indeterminates, and $e := (e_1, \ldots, e_N) \in \mathbb{N}^N$ a vector with positive coordinates. Consider the formal power series ring

$$B := A[[X]]$$

and the filtration

$$F_B^d := \sum_{i \in \mathbb{N}^r} F_A^{d - <i,e>} \cdot X^i \cdot B$$

generated over $F_A \cdot B$ by the family X such that $\operatorname{ord} X_i \ge e_i$. It is easy to see (cf 3.9) that the order of X_i with respect to this filtration is equal to e_i for every i. The power series ring B equipped with this filtration will be denoted

$$A[[X \,|\, \operatorname{ord} X = e]] = A[[X_1, \ldots, X_N \,|\, \operatorname{ord} X_1 = e_1, \ldots, \operatorname{ord} X_N = e_N]]$$

In case all coordinates of e are equal to 1, the corresponding filtered ring will be simply denoted

$$A[[X]] := A[[X \,|\, \operatorname{ord} X_i = 1]].$$

In the general case,

$$G(A[[X \,|\, \operatorname{ord} X = e]])(d) = \bigoplus_{<i,e> \leq d} G(A)(d - < i, e >) \cdot X^i$$

Hence $G(A[[X \,|\, \operatorname{ord} X = e]]) = G(A)[X]$ is the polynomial ring over $G(A)$ with X_i being a variable of degree e_i.

6.7. The Hilbert series of a power series extension

Let (A, m) be a filtered local ring, $X = \{X_1, \ldots, X_N\}$ a finite set of indeterminates, and $e := (e_1, \ldots, e_N) \in \mathbb{N}^N$ a vector with positive coordinates. Then

$$\mathrm{H}^0_{A[[X \,|\, \operatorname{ord} X = e]]} = \mathrm{H}^0_A \cdot \prod_{i=1}^{N} \frac{1}{1 - T^{e_i}}.$$

Proof. Define $B := A[[X \,|\, \operatorname{ord} X = e]]$. We may assume that $N = 1$. From the exact sequence of homomorphisms of graded rings

$$0 \to G(B)[-e_1] \xrightarrow{X_1} G(B) \to G(B)/X_1 \, G(B) \to 0$$

and the fact that $G(B)/X_1 \, G(B) \cong G(A)$ one sees,

$$\mathrm{H}_{G(B)} \cdot T^{e_1} - \mathrm{H}_{G(B)} + \mathrm{H}_{G(A)} = 0,$$

i.e., $(1 - T^{e_1}) \cdot \mathrm{H}_{G(B)} = \mathrm{H}_{G(A)}$, which is just the claim.

6.8. The Hilbert series of a factor ring

Let (A, m) be a filtered local ring and $x \in m$ an element of order $\geq \mu$. Then

(i) $\mathrm{H}^1_{A/xA} \geq (1 - T^\mu) \cdot \mathrm{H}^1_A$

(ii) Equality holds in (i) if and only if the following equivalent conditions are satisfied.

    a) $F^d_A : x = F^{d-\mu}_A$ for every $d \in \mathbb{N}$.

    b) $\operatorname{in}(x)$ is a regular element of degree μ in $G(A)$.

If the conditions of (ii) are satisfied, the initial ideal of xA is generated by $\operatorname{in}(x)$,

$$\operatorname{in}(xA) = \operatorname{in}(x) \cdot G(A).$$

Proof. (i). Since x has order at least μ, $F_A^{d+1-\mu} \subseteq F_A^{d+1} : x$. Hence there is a surjection

$$A/F_A^{d+1-\mu} \to A/(F_A^{d+1} : x) \cong (xA + F_A^{d+1})/F_A^{d+1}$$

where the isomorphism on the right is induced by multiplication with x. Therefore,

$$\begin{aligned}
\mathrm{II}_{A/xA}^1(d) &= L(A/xA + F_A^{d+1}) \\
&= L(A/F_A^{d+1}) - L(xA + F_A^{d+1}/F_A^{d+1}) \\
&\geq \mathrm{H}_A^1(d) - \mathrm{H}_A^1(d - \mu),
\end{aligned}$$

i.e., $\mathrm{H}_{A/xA}^1 \geq \mathrm{H}_A^1 - T^\mu \cdot \mathrm{H}_A^1$ as claimed.

(ii). Equality holds in (i) if and only if one has equality in the above estimation, i.e., if and only if $F_A^{d+1-\mu} = F_A^{d+1} : x$ for every d. This is condition a) of (ii). Now assume that condition a) is satisfied. Then

$$F_A^d \cap (xA) = F_A^{d-\mu} x \quad \text{for every } d$$

(multiply by x the given identities), in particular, $x \in xA = F_A^\mu \cap (xA)$, so that x has order μ in A, i.e., $\mathrm{in}(x)$ degree μ in $G(A)$. Assume that

$$\mathrm{in}(x) \cdot \mathrm{in}(x') = 0, \mathrm{in}(x') \neq 0$$

for some element $x' \in A$ of order μ'. Then $xx' \in F_A^{\mu+\mu'+1}$ hence $x' \in F_A^{\mu+\mu'+1} :$ $x = F_A^{\mu'+1}$, hence $\mathrm{ord}\, x' > \mu'$, contrary to the definition of μ'. Thus $\mathrm{in}(x)$ cannot be a zero divisor. Conversely, assume that (ii)b) is satisfied. We have yet to show that then equality holds in (i) and that the assertion concerning the initial ideal of xA is true. The assumption implies that multiplication by $\mathrm{in}(x)$ defines an exact sequence of graded modules over $G(A)$,

$$0 \to G(A)[-\mu] \xrightarrow{\mathrm{in}(x)} G(A) \to G(A)/(\mathrm{in}(x)) \to 0.$$

Therefore, since $(\mathrm{in}(x)) \subseteq \mathrm{in}(xA)$,

$$\begin{aligned}
\mathrm{H}_{A/xA}^1 &= \mathrm{H}_{G(A)/\mathrm{in}(xA)}^1 \\
&\leq \mathrm{H}_{G(A)/(\mathrm{in}(x))}^1 \\
&= \mathrm{H}_{G(A)}^1 - \mathrm{H}_{G(A)}^1 \cdot T^\mu \\
&= (1 - T^\mu) \cdot \mathrm{H}_{G(A)}^1 .
\end{aligned}$$

According to (i) equality holds everywhere. But then, $(\mathrm{in}(x)) = \mathrm{in}(xA)$. This completes the proof.

6.9. Case of a power series algebra modulo a regular sequence

Let (A, m) be a filtered local ring, $X = \{X_1, \ldots, X_N\}$ a finite set of indeterminates, $e := (e_1, \ldots, e_N) \in \mathbb{N}^N$ a vector with positive coordinates, and

$f_1, \ldots, f_s \in A[[X | \operatorname{ord} X = e]]$ power series with positive orders $\mu_1, \ldots, \mu_s$, respectively. Define

$$B := A[[X | \operatorname{ord} X = e]]/(f_1, \ldots, f_s).$$

Then

(i) $H_B^s \geq H_A^s \cdot \prod_{j=1}^{s} (1 - T^{\mu_j}) \cdot \prod_{i=1}^{r} (1 - T^{e_i})^{-1}$

(ii) Equality holds in (i) if and only if the $\operatorname{in}(f_i)$'s form a regular sequence of the graded algebra $G(A[[X | \operatorname{ord} X = e]])$.

Proof. Define

$$R := A[[X | \operatorname{ord} X = e]] \quad \text{and} \quad B_j := R/(f_1, \ldots, f_j).$$

Then, by 6.8,

$$(1) \qquad\qquad H_{B_{j+1}}^{j+1} \geq \frac{(1 - T^{\mu_{j+1}})}{1 - T} \cdot H_{B_j}^j,$$

hence by 6.7,

$$H_B^s \geq H_R^0 \cdot (1 - T)^{-s} \cdot \prod_{j=1}^{s} (1 - T^{\mu_j}) = H_A^s \cdot \prod_{j=1}^{s} (1 - T^{\mu_j}) \cdot \prod_{i=1}^{r} (1 - T^{e_i})^{-1}$$

Note that the quotients $(1 - T^{\mu_{j+1}})/(1 - T)$ are polynomials with non-negative coefficients. Equality holds in (i) if and only if equality holds in all the estimations (1). But this means, by 6.8, that the initial forms of the generators f_j form a regular sequence in $G(A[[X | \operatorname{ord} X = e]])$.

6.10. Hilbert series of a local extension

Let $f : (A, m) \to (B, n)$ be a homomorphism of filtered local rings with special fibre $\bar{B} = B/mB$. Then

$$H_B^1 \leq H_A^1 \cdot H_{\bar{B}}^0$$

Proof. From the basic exact sequence 3.14

$$0 \to D(d, i) \to \mathrm{GF}_A(B)(i) \otimes_B B(d - i) \to \mathrm{GF}_A(B(d))(i) \to 0$$

we obtain

$$(1) \qquad \begin{cases} H_B^1(d) & = L_B(B(d)) = \sum_{i=0}^{d} L_B(\mathrm{GF}_A(B(d))(i)) \\ & = \sum_{i=0}^{d} \left(L_B(\mathrm{GF}_A(B)(i) \otimes_B B(d - i)) - L_B(D(d, i)) \right) \end{cases}$$

hence

$$H^1_B(d) \leq \sum_{i=0}^d L_B(G_{F_A}(B)(i) \otimes_B B(d-i)).$$

Consider the canonical surjection $G_{F_A}(A) \otimes_A B \to G_{F_A}(B)$. Tensor with $B(d-i)$ over B to get a surjective homomorphism

$$G_{F_A}(A) \otimes_A B(d-i) \to G_{F_A}(B) \otimes_B B(d-i)$$

that allows further estimation:

$$H^1_B(d) \leq \sum_{i=0}^d L_B(G_{F_A}(A)(i) \otimes_A B(d-i))$$

$$\leq \sum_{i=0}^d L_A(G_{F_A}(A)(i)) L_A(\bar{B}(d-i))$$

$$= (H_A \cdot H^1_{\bar{B}})(d).$$

See 2.8 for the second inequality. We have proved

$$H^1_B \leq H_A \cdot H^1_{\bar{B}} = (1-T)^{-1} H_A \cdot H_{\bar{B}} = H^1_A \cdot H^0_{\bar{B}}.$$

6.11. The situation when H^1 is replaced by H^0

One cannot expect to improve the above estimation such that

$$H^0_B \leq H^0_A \, H^0_{B/mB}.$$

To see this, let K be a field and x, y, z, X, Y be indeterminates over K. Define

$$A := K[[x,y,z]]/(x,y,z)^2$$
$$B := A[X,Y]/(X^2-x, XY-y, Y^2-z) \quad (\cong K[X,Y]/(X,Y)^4)$$

Equip the local rings (A,m) and (B,n) with their respective natural filtrations and consider the natural homomorphism $A \to B$. Then

$$H^0_A = 1+3T, \quad H^0_{B/mB} = 1+2T, \quad H^0_B = 1+2T+3T^2+4T^3,$$

hence $H^0_A \cdot H^0_{B/mB} = 1+5T+6T^2$ and the inequality above doesn't hold. Even for flat local homomorphisms $f : (A,m) \to (B,n)$ the strengthened inequality is generally wrong. The following (probably easiest) example is due to C. Lech. Let $R := K[[X]]$ be a power series ring over the field K in one indeterminate X and

$$f' : R \to R, \quad p(X) \mapsto p(X^2)$$

the continuous K-algebra homomorphism mapping X to X^2. Define

$$A := R/(X^2), \quad B := R/(X^4), \text{ and } f := f' \otimes_R R/(X^2) : A \to B.$$

Then the special fibre of f is $\bar{B} = R/(X^2)$ and the Hilbert series satisfy

$$\mathrm{H}_A = 1 + T, \quad \mathrm{H}_B = 1 + T + T^2 + T^3, \quad \mathrm{H}_{\bar{B}} = 1 + T.$$

In particular, H_B is not $\leq 1 + 2T + T^2 = \mathrm{H}_A \cdot \mathrm{H}_{\bar{B}}$.

Out next aim is to derive conditions ensuring equality in the estimation of 6.10. We shall first treat the case of graded rings.

6.12. Flat homomorphisms of graded rings

Let $f : G \to G'$ be a graded ring homomorphism satisfying the following conditions.

1. G and G' have finite Hilbert functions.
2. $G(0)$ is a local ring with maximal ideal, say, $m(0)$.

Write $m := m(0) + G^+$ for the (only) maximal homogeneous ideal of G. Then

(i) $\mathrm{H}_{G'}^0 \leq \mathrm{H}_G^0 \cdot \mathrm{H}_{G'/mG'}^0$

(ii) The following conditions are equivalent.

    (a) Equality holds in (i).

    (b) G' is flat over G.

    (c) G'/G^+G' is flat over $G/G^+ (\cong G(0))$, and there is a graded $G(0)$-module section $s : G'/G^+G' \to G'$ of the canonical homomorphism $G' \to G'/G^+G'$ such that the induced surjection of G-modules

$$(1) \qquad G \otimes_{G(0)} G'/G^+G' \to G', \quad g \otimes m \mapsto f(g) \cdot s(m)$$

    is an isomorphism (in which case (1) is an isomorphism for every graded section s).

We first prove a lemma.

Lemma. Let $f : G \to G'$ be a graded ring homomorphism as in the above statement, i.e., such that

1. G and G' have finite Hilbert functions.
2. $G(0)$ is a local ring with maximal ideal $m(0)$.

Write $m := m(0) + G^+$. Then

(i) $\mathrm{H}_{G'}^0 \leq \mathrm{H}_G^0 \cdot \mathrm{H}_{G'/mG'}^0$

(ii) Suppose equality holds in (i). Then this equality is preserved under the substitution $f \mapsto f \otimes_G G/I$, i.e.,

$$\mathrm{H}_{G'/IG'}^0 = \mathrm{H}_{G/I}^0 \cdot \mathrm{H}_{G'/mG'}^0$$

for every proper homogeneous ideal $I \subseteq G$. Moreover, given a homogeneous non-unit $g \in G$, multiplication by g defines short exact sequences

$$0 \to G_2[-\deg g] \overset{g}{\to} G \to G_1 \to 0, \quad G_1 := G/gG, \; G_2 := G/(0_G : g)G,$$

$$0 \to G_2'[-\deg g] \overset{g}{\to} G' \to G_1' \to 0. \quad G_1' := G'/gG', \; G_2' := G'/(0_G : g)G'.$$

(iii) Assume that there is some homogeneous non-unit $g \in G - 0$ such that the identity of (ii) holds for $I = (g)$ and $I = (0_G : g)$. Then this identity holds for every proper homogeneous ideal $I \subseteq G$.

Proof of the lemma. Let $g \in G$ be an arbitrary homogeneous non-unit of degree, say, e. Then the first of the above two sequences is exact by definition. The second one is obtained from the first tensoring with G', so it can be non-exact only at $G_2'[-e]$. Therefore,

$$\mathrm{H}_G^0 = \mathrm{H}_{G_1}^0 + \mathrm{H}_{G_2}^0 \cdot T^e \quad \text{and} \quad \mathrm{H}_{G'}^0 \leq \mathrm{H}_{G_1'}^0 + \mathrm{H}_{G_2'}^0 \cdot T^e,$$

and equality in the estimation on the right is equivalent to short exactness of the second sequence. The claim of the lemma will follow now by an induction argument.

In order to prove the inequality of (i), define

$$I_N := \overset{\infty}{\underset{d=N+1}{\oplus}} G(d), \quad N = 1, 2, 3, \dots$$

and note that the first N coefficients of the involved Hilbert series remain unchanged when f is replaced by $f \otimes G/I_N$. So we may assume that $G(d) = 0$ for large d. Then the ideal G^+ is nilpotent and G is a local ring with finite length. We shall prove the inequality of (i) by induction on the length

$$L := L(G).$$

In case $L = 1$, G is a field, $\mathrm{H}_G = 1$, $m = 0$, $\mathrm{H}_{G'} = \mathrm{H}_{G'/mG'}$, and the claim is trivial. In case $L > 1$, take some non-zero homogeneous non-unit $0 \neq g \in G$. Then, $(0_G : g)$ is a non-trivial ideal of G (since G is Artinian), hence the induction hypothesis applies to $f \otimes_G G_1$ and $f \otimes_G G_2$. Therefore,

$$\mathrm{H}_{G'}^0 \leq \mathrm{H}_{G_1'}^0 + \mathrm{H}_{G_2'}^0 \cdot T^e$$
$$\leq (\mathrm{H}_{G_1}^0 + \mathrm{H}_{G_2}^0 \cdot T^e) \cdot \mathrm{H}_{G'/mG'}^0$$
$$= \mathrm{H}_G^0 \cdot \mathrm{H}_{G'/mG'}^0$$

This proves the first part of the lemma.

Note that in view of assertion (i) of the lemma, the last estimation is now valid also for the general (i. e., non-Artinian) case. If the identity of (ii) is true for $I = (g)$ and $I = (0_G : g)$, then equality holds everywhere in this estimation, i.e., the identity of (ii) is also true for $I = (0)$. Part (ii) of the lemma claims now, that the identity is true for arbitrary I. This reduces the proof of assertion (iii) to the proof of (ii).

So assume equality holds in (i). Then equality holds everywhere in the above estimation. In particular,

$$\mathrm{H}^0_{G'} = \mathrm{H}^0_{G'_1} + \mathrm{H}^0_{G'_2} \cdot T^e \quad \text{and} \quad \mathrm{H}^0_{G'_1} = \mathrm{H}^0_{G_1} \cdot \mathrm{H}^0_{G'/mG'} \, .$$

The identity on the left implies that the second sequence of (ii) is short exact, as required, and the right hand side identity means that the identity of (ii) holds for principal ideals $I = (g)$. Repeated application of the latter result gives the claim for I finitely generated. If I is arbitrary, one can find a finite number of elements generating I up to homogeneous elements above a fixed degree N (since G has finite Hilbert function) so that we have the identity modulo T^N. Choosing N larger and larger, we get the claim for all coefficients of the power series involved.

Proof of 6.12. Assertion (i) of 6.12 is identical with assertion (i) of the lemma. So all we have to prove is the equivalence of the conditions in (ii).

Note that the homomorphism (1) of condition (ii)(c) is surjective whenever it exists. For, if there is a section s as in (ii)(c), then

$$G' = \mathrm{Im}(s) + G^+ G'$$

hence $G' = f(G) \cdot \mathrm{Im}(s) + G^+ G'$. The graded Nakayama Lemma 2.1 now implies $G' = f(G) \cdot \mathrm{Im}(s)$, i.e., the mapping (1) is surjective.

(b)⇒(a). Since base change doesn't effect the flatness assumption, we may assume that $G(d) = 0$ for large d, hence $L := L(G) < \infty$. In case $L = 1$ we have $\mathrm{H}_G = 1$, $m = 0$, $\mathrm{H}_{G'} = \mathrm{H}_{G'/mG'}$, i.e., the claim of (a) is trivially true. So let $L > 1$. Inductively we may assume assertion (a) is true for every homomorphism $f \otimes_G G/I$ and every non-zero homogeneous ideal I. But then by part (iii) of the lemma, (a) must also be true for f itself.

(c)⇒(b). By assumption, $G'/G^+ G'$ is $G(0)$-flat, hence $G' \cong G \otimes_{G(0)} G'/G^+ G'$ is G-flat.

(a)⇒(c). It will be sufficient to show that for every positive integer N the following assertion is true.

Claim. $(G'/G^+ G')(d)$ is flat over $G(0)$ for $d = 0,\dots,N$ and for every family of maps $(s_d)_{d=0,\dots,N}$ where $s_d : (G'/G^+ G')(d) \to G'(d)$ is a section of the degree d part of the canonical map $G' \to G'/G^+ G'$, the induced homomorphism

$$\bigoplus_{i+j=N} G(i) \otimes_{G(0)} (G'/G^+ G')(j) \to G'(N)$$

is bijective.

Note that the claim implies, since $G(0)$ is Artinian, that $(G'/G^+G')(d)$ is free over $G(0)$ for $d = 0,1,2,\ldots,N$, so that the existence of the sections s_d is no problem. To prove the claim, we may assume that $G(d) = 0$ for large d (in view of the lemma), which reduces the proof to the case that $L := L(G)$ is finite. If $L = 1$, the ring $G = G(0)$ is a field, G^+ is the zero ideal, and the (ii)c) is trivially satisfied. Let $L > 1$. Choose some non-zero homogeneous non-unit $g \in G$ of degree, say, e. Then we may inductively assume that the assertions of 6.12(ii) are valid for the homomorphisms

$$f_1 := f \otimes_G G/gG : G_1 \to G_1' \text{ and } f_2 := f \otimes_G G/(0_G : g)G : G_2 \to G_2'.$$

In view of the second part of the lemma, condition (ii)(a) of 6.12 is satisfied for these homomorphisms, hence so are the other conditions. In particular, the homomorphism f_1 is flat. Further we know from the lemma that there are two short exact sequences. Since $G_2 = G/(0_G : g)G \cong gG$ and $G_2' = G/(0_G : g)G \otimes_G G' \cong gG \otimes_G G'$, the exactness of the second one implies that the canonical homomorphism $gG \otimes_G G' \to gG'$ is bijective. Hence f is flat by the local flatness criterion ([Ma86], Th. 22.3(3)) and G'/G^+G' is $G(0)$-flat, i.e., the first part of the claim is true.

To prove the remaining part, choose a family of homomorphisms

$$s_d : (G'/G^+G')(d) \to G'(d)$$

$(d = 0,1,2,\ldots)$ defining a graded $G(0)$-module section $s : G'/G^+G' \to G'$ of the canonical homomorphism $G' \to G'/G^+G'$. This is possible, since G'/G^+G' is $G(0)$-flat (hence free). The map $G \otimes_{G(0)} G'/G^+G' \to G'$ of condition (ii)(c) defined by the section s can be included as follows into a commutative diagram.

$$
\begin{array}{ccc}
0 & & 0 \\
\downarrow & & \downarrow \\
G_2[-e] \otimes_{G(0)} G'/G^+G' & \to & G_2'[-e] \\
\downarrow & & \downarrow \\
G \otimes_{G(0)} G'/G^+G' & \to & G' \\
\downarrow & & \downarrow \\
G_1 \otimes_{G(0)} G'/G^+G' & \to & G_1' \\
\downarrow & & \downarrow \\
0 & & 0
\end{array}
$$

The upper row of the diagram is obtained from the middle one by tensoring with G_2 over G, and the lower by tensoring with G_1. The right hand side column is defined by multiplication in G' with the element g, and is exact by the lemma. The left hand side column is obtained from the short exact sequence

$$0 \to G_2[-e] \xrightarrow{g} G \to G_1 \to 0$$

tensoring with the flat $G(0)$-module G'/G^+G' hence is also exact. By induction hypothesis, the upper and the lower horizontal maps are bijective. Note that

gG and $(0 : g)G$ are non-trivial ideals in G, the first by assumption and the second since G is Artinian. By the five lemma the homomorphism in the middle is bijective. This proves (ii)(c) of 6.12 and completes the proof of the theorem.

6.13. Hilbert series of a local extension and tangential flatness

Let $f : (A, m) \to (B, n)$ be a homomorphism of filtered local rings with special fibre $\bar{B} = B/mB$. Then the following conditions are equivalent.

(i) $H_B^1 = H_A^1 \, H_{\bar{B}}$

(ii) $H_{G(B)}^1 = H_{G(A)}^1 \, H_{G(B)/m^* \, G(B)}^0$, $\quad m^* := m\,G(A) + G^+(A)$

(iii) $f : (A, m) \to (B, n)$ is tangentially flat.

(iv) $G(B/F_A^1 B)$ is flat over A/F_A^1 and there is a graded A-module section

$$ s : G(B/F_A^1 B) \to G(B) $$

of the canonical surjection $G(B) \to G(B/F_A^1 B)$ such that the induced homomorphism of graded $G(A)$-modules

$$ G(A) \otimes_A G(B/F_A^1 B) \to G(B) $$

is injective (in which case it is an isomorphism for every graded section s).

Proof. (i)$\Rightarrow$(iii). Analyzing the proof 6.10 of the inequality

$$ H_B^1 \leq H_A^1 \, H_{\bar{B}} $$

we see that condition (i) implies that for arbitrary d and i with $i \leq d$,

$$ (1) \qquad L_B(G_{F_A}(A)(i) \otimes_A B(d-i)) = H_A^0(i) \cdot H_{\bar{B}}^1(d-i) $$

and the composition of natural surjections

$$ G_{F_A}(A) \otimes_A B(d-i) \to G_{F_A}(B)(i) \otimes_B B(d-i) \to G_{F_A}(B(d))(i) $$

is an isomorphism. To prove that $f : (A, m) \to (B, n)$ is tangentially flat, it will be sufficient (by 3.15) to show that (1) implies that $G(B/F_A^1 B)$ is flat over A/F_A^1. Define

$$ A_0 := A/F_A^1 \quad \text{and} \quad B_0 := B/F_A^1 B. $$

In view of the exact sequences

$$ 0 \to G(B_0)(d) \to B_0(d) \to B_0(d-1) \to 0 $$

it is enough to show that $B_0(d)$ is flat over A_0 for every d. Since the rings involved are Artinian, the latter is equivalent (by 2.8(iii)) to the identities

$$ L(B_0(d)) = L(A_0)L(B_0(d)/mB_0(d)). $$

But this is just (1) with $i = 0$. Note that

$$L_B(\mathrm{G}_{F_A}(A)(0) \otimes_A B(d)) = L(B_0(d)),$$
$$\mathrm{H}_A^0(0) = L(A_0),$$
$$\mathrm{H}_{\bar{B}}^1(d) = L(B_0(d)/mB_0(d)).$$

(iii)$\Rightarrow$(ii). The associated graded homomorphism $G(A) \to G(B)$ is flat by assumption, hence by 6.12

$$\mathrm{H}_{G(B)}^1 = \mathrm{H}_{G(A)}^1 \, \mathrm{H}_{G(B)/m^{\bullet} \, G(B)}^0 .$$

(ii)$\Rightarrow$(i). The ideal $m^* = m \, G(A) + G^+(A)$ is maximal in $G(A)$ and annihilates the graded ring
$$G(B/mB) = G(B)/\mathrm{in}(mB).$$

So there is a surjection $G(B)/m^* \, G(B) \to G(B/mB)$ which yields the estimation

$$\mathrm{H}_B^1 = \mathrm{H}_{G(B)}^1 = \mathrm{H}_{G(A)}^1 \cdot \mathrm{H}_{G(B)/m^{\bullet} \, G(B)}^0 \geq \mathrm{H}_{G(A)}^1 \cdot \mathrm{H}_{G(B/mB)}^0 = \mathrm{H}_A^1 \cdot \mathrm{H}_{B/mB}^0$$

Since the converse inequality is always true (by 6.10), this gives the claim of (i).

(iv)$\Rightarrow$(iii). Since $G(B/F_A^1 B)$ is flat over A/F_A^1, the tensor product $G(B) \cong G(A) \otimes_A G(B/F_A^1 B)$ is flat over $G(A)$.

(iii)$\Rightarrow$(iv). Flatness of $G(A) \to G(B)$ implies that the conditions of 6.12 are satisfied. To prove the claim, it will be sufficient to show that

$$G(B)/ \, G^+(A) \, G(B) \cong G(B/F_A^1 B).$$

By 3.8 this is a consequence of the fact that $G^+(A) = \mathrm{in}(F_A^1)$.

6.14. The case of Artinian rings

Let $f : (A, m) \to (B, n)$ be a homomorphism of filtered local rings with special fibre $\bar{B} = B/mB$. Assume that

$$L(A) < \infty \text{ and } L(B) < \infty.$$

Then the following conditions are equivalent.

(i) f is tangentially flat.

(ii) $\mathrm{H}_B^0 \geq \mathrm{H}_A^0 \, \mathrm{H}_{\bar{B}}^0$.

(iii) $\mathrm{H}_B^0 \leq \mathrm{H}_A^0 \, \mathrm{H}_{\bar{B}}^0$ and B is flat over A.

Proof. Implications (i)$\Rightarrow$(ii) and (i)$\Rightarrow$(iii) are known (see 6.13 and 4.2). Implication (ii)$\Rightarrow$(i) is also easy, for, $\mathrm{H}_B^0 \geq \mathrm{H}_A^0 \, \mathrm{H}_{\bar{B}}^0$ implies $\mathrm{H}_B^1 \geq \mathrm{H}_A^1 \, \mathrm{H}_{\bar{B}}^0$ (multiply by $(1 - T)^{-1} = 1 + T + T^2 + \ldots$) hence by 6.10, $\mathrm{H}_B^1 = \mathrm{H}_A^1 \, \mathrm{H}_{\bar{B}}^0$ which we know to be equivalent to tangential flatness (by 6.13). So (iii)$\Rightarrow$(i) is the only implication yet to be proved. By assumption,

$$H^0_B(d) \leq \sum_{i+j=d} H^0_A(i)\, H_{\bar B}(j).$$

We have to show, equality holds for every d. Assume the inequality is strict for at least one d. Then, taking the sum over all d, we get a strict inequality

$$L(B) = \sum_{d=0}^{\infty} H^0_B(d) < \sum_{d=0}^{\infty} \sum_{i+j=d} H^0_A(i)\, H_{\bar B}(j) = L(A)L(\bar B),$$

which is impossible (by 2.8), since B is flat over A by assumption.

Remark

We conclude the section with two results on the composition of tangentially flat morphisms, which are due to late Christer Lech. The author learned about them in a private communication.

6.15. Composition and tangential flatness

Let $f : (A, m) \to (B, n)$ and $g : (B, n) \to (C, p)$ be a homomorphisms of filtered local rings such that the following conditions are satisfied.

1. $gf : A \to B \to C$ is tangentially flat.
2. $g \otimes_A A/m : B/mB \to C/mC$ is tangentially flat.

Then both $f : A \to B$ and $g : B \to C$ are tangentially flat.

Proof. By the second condition (cf 6.13),

$$H^0_{C/mC} = H^0_{B/mB} \cdot H^0_{C/nC}.$$

Therefore by 6.10,

$$H^1_C \leq H^1_B \cdot H^0_{C/nC} \leq H^1_A \cdot H^0_{B/mB} \cdot H^0_{C/nC} = H^1_A \cdot H^0_{C/mC}.$$

By the first condition, equality holds everywhere in the above estimation. Thus,

$$H^1_C = H^1_B \cdot H^0_{C/nC} \quad \text{and} \quad H^1_B = H^1_A \cdot H^0_{B/mB},$$

i.e., f and g are tangentially flat as claimed.

6.16. Cartesian diagrams and tangential flatness

Let

$$
\begin{array}{ccc}
(A, m) & \xrightarrow{\ f\ } & (A', m') \\
h \downarrow & & \downarrow h' \\
(B, n) & \xrightarrow{\ g\ } & (B', n')
\end{array}
$$

be a commutative diagram of filtered local rings such that the following conditions are satisfied.

1. For each pair of opposite homomorphisms the associated fibres have equal Hilbert series (i.e., $H_f^0 = H_g^0$ and $H_h^0 = H_{h'}^0$).

2. There is a pair of two consecutive tangentially flat homomorphisms (say, h and g).

Then all homomorphisms of the diagram are tangentially flat.

Proof. Assume h and g are tangentially flat. Then we have the following estimation (by 6.10 and 6.13).

$$
\begin{aligned}
H_{A'}^1 \cdot H_h^0 &\leq H_A^1 \cdot H_f^0 \cdot H_h^0 \\
&= H_B^1 \cdot H_f^0 \quad \text{(since } h \text{ is tangentially flat)} \\
&= H_B^1 \cdot H_g^0 \quad \text{(by condition 1)} \\
&= H_{B'}^1 \quad \text{(since } g \text{ is tangentially flat)} \\
&\leq H_{A'}^1 \cdot H_{h'}^0 \\
&= H_{A'}^1 \cdot H_h^0 \quad \text{(by condition 1)}
\end{aligned}
$$

We see that equality must hold everywhere in this estimation. Therefore, $H_{A'}^1 = H_A^1 \cdot H_f^0$ and $H_{B'}^1 = H_{A'}^1 \cdot H_{h'}^0$, i.e., f and h' are tangentially flat (by 6.13).

7. Flatifying filtrations

Remark

In this chapter we want to prove that, given a flat homomorphism

$$f : (A, m) \to (B, n)$$

of filtered local rings, the filtration F_A can be enlarged in such a way that f becomes tangentially flat. Moreover, it will turn out that among the possible enlargements there is a unique minimal filtration with this property. The main problem in proving these results is to show that the enlarged filtration is an Artin-Rees filtration. Therefore most theorems in this chapter will deal with prefiltrations (see 3.11) rather than filtrations. A side result, which is certainly of its own interest, implies that for every flat local homomorphism $f : (A, m) \to (B, n)$ of naturally filtered local rings there is a tangentially flat finite covering (see 7.8). In particular, Lech's inequality $H_{A'}^0 \leq H_{B'}^0$ holds for a finite covering $f' : (A', m') \to (B', n')$ of f.

Despite of the fact that we are dealing with prefiltrations, the proofs in this section will sometimes rely on theorems formulated for filtrations only. However, this will happen exclusively in the complete case, where every cofinite prefiltration F of A becomes a filtration when the ring A is replaced by the factor ring $A/\cap F$ (see 1.15(iv)). We will use this fact without further reference.

7.1. Flatifying filtrations

Let $f : (A, m) \to (B, n)$ be a homomorphism of filtered local rings and F a ring prefiltration (see 3.11) of A. Then F is called a *flatifying prefiltration* for f, if the following conditions are satisfied.

1. $F_A \subseteq F$ and $F^1 \subseteq m$.

2. $f : (A, F) \to (B, F_B + f_* F)$ is tangentially flat.

A *flatifying filtration* is a flatifying prefiltration which is a ring filtration in the sense of 3.11.

7.2. An Example of a flatifying prefiltration

For every homomorphism $f : (A, m) \to (B, n)$ of filtered local rings, the prefiltration

$$F^d := \begin{cases} A & \text{if } d = 0 \\ m & \text{otherwise} \end{cases}$$

is always a flatifying prefiltration. For, $G_F(A) \cong A/m$ is a field.

7.3. Intersection of flatifying prefiltrations

Let $f : (A, m) \rightarrow (B, n)$ be a homomorphism of filtered local rings and let $(F_\gamma)_{\gamma \in \Gamma}$ be a family of flatifying prefiltrations for $f : (A, m) \rightarrow (B, n)$. Then the intersection

$$F := \bigcap_{\gamma \in \Gamma} F_\gamma$$

is a flatifying prefiltration for f.

Proof. Choose some F_B-distinguished basis (b, d) of B over A. Then it will be sufficient to show that the following two conditions are satisfied (see 2.6).

(a) $B \subseteq \sum_{\lambda \in \Lambda} A \cdot b(\lambda) + (F_B + f_* F)^d$ for every $d \in \mathbb{N}$.

(b) $\sum_{\lambda \in \Lambda} a_\lambda b(\lambda) \in (F_B + f_* F)^d$ implies $a_\lambda \in F^{d-d(\lambda)}$ for every $\lambda \in \Lambda$

In fact, these conditions mean that the elements $b(\lambda)$ represent a free generating set for the associated graded ring (as a module over the graded base ring). However, this is equivalent to tangential flatness by 4.2.

For arbitrary $\gamma \in \Gamma$, the prefiltration induced by $F_B + f_* F_\gamma$ on the special fibre B/mB of f is independent upon γ (and equal to the filtration induced by F_B). Since F_γ is a flatifying prefiltration for f, $G_{F_B+f_*F_\gamma}(B)$ is $G_{F_\gamma}(A)$-flat. So the two conditions (a) and (b) above are satisfied if F is replaced by any of the prefiltrations F_γ. In order to prove condition (b), assume that

$$\sum_{\lambda \in \Lambda} a_\lambda b(\lambda) \in (F_B + f_* F)^d.$$

Since $F \subseteq F_\gamma$, this relation is preserved when F is replaced by F_γ. Therefore, since (b) holds for F_γ,

$$a_\lambda \in F_\gamma^{d-d(\lambda)} \text{ for arbitrary } \lambda \in \Lambda \text{ and arbitrary } \gamma \in \Gamma.$$

But then $a_\lambda \in \bigcap_{\gamma \in \Gamma} F_\gamma^{d-d(\lambda)} = F^{d-d(\lambda)}$. We have proved, condition (b) is satisfied. As for condition (a) note that, since (b, d) is F_B-distinguished,

$$F_B^d \subseteq \sum_{\lambda \in \Lambda} A \cdot b(\lambda) + F_B^{d+1} + mB.$$

Therefore

$$B \subseteq \sum_{\lambda \in \Lambda} A \cdot b(\lambda) + F_B^d + mB$$

for $d = 0,1,2,\ldots$. Iterating the latter inclusion, the maximal ideal m of A may be replaced by an arbitrary power m^i. Since F_B is cofinite, $m^i B \subseteq F_B^d$ for large i, i.e.,

$$B \subseteq \sum_{\lambda \in \Lambda} A \cdot b(\lambda) + F_B^d$$

for every $d \in \mathbb{N}$. Replace F_B^d by the larger ideal $(F_B + f_* F)^d$ to obtain inclusion (a).

7.4. The minimal flatifying prefiltration

Let $f : (A, m) \to (B, n)$ be a homomorphism of filtered local rings. Then the *minimal flatifying prefiltration* for f, denoted

$$F(f),$$

is by definition the intersection of all flatifying prefiltrations for f.

7.5. Minimal flatifying prefiltrations and structure constants

Let $f : (A, m) \to (B, n)$ be a flat homomorphism of complete filtered local rings, and (b, d) a distinguished basis for f. Assume that F_B is F_A-minimal with respect to (b, d) (see 5.4). Moreover, let

(1) $$(a_{\alpha\beta}^{\gamma})_{\alpha,\beta,\gamma \in \Lambda}$$

be the family of structure constants with respect to (b, d). Then the minimal flatifying prefiltration for f (see 7.4),

$$F = F(f),$$

is just the ring filtration generated over F_A (see 1.11) by the family (1) such that

$$\mathrm{ord}(a_{\alpha\beta}^{\gamma}) \geq d(\alpha) + d(\beta) - d(\gamma).$$

Proof. Let F' be the ring prefiltration generated over F_A by the family (1) such that

$$\mathrm{ord}(a_{\alpha\beta}^{\gamma}) \geq d(\alpha) + d(\beta) - d(\gamma).$$

In order to show that F is a flatifying prefiltration, we only have to prove that $F_B + f_* F$ is F-minimal with respect to (b, d) (see 5.8). So let's calculate the

d-th filtration ideal of $F_B + f_* F$. Since F_B is F_A-minimal and $F_A \subseteq F$,

$$(F_B + f_* F)^d = \sum_j F^{d-j} \cdot F_B^j$$

$$= \sum_j F^{d-j} \cdot \sum_{(i,\lambda) \in \mathbb{N}^n \times \Lambda^n} F_A^{j - \langle i, \mathrm{d}(\lambda) \rangle} \cdot B \cdot \mathrm{b}(\lambda)^i$$

$$= \sum_{(i,\lambda) \in \mathbb{N}^n \times \Lambda^n} F^{d - \langle i, \mathrm{d}(\lambda) \rangle} \cdot B \cdot \mathrm{b}(\lambda)^i,$$

i.e., $F_B + f_* F$ is F-minimal (see 5.4). We have proved, F is a flatifying pre-filtration for f. We have yet to show that F is contained in every flatifying prefiltration F' for f. As we know, the flatifying prefiltrations F' for f satisfy (by 5.8),

$$\mathrm{ord}_{F'}\, a_{\alpha\beta}^\gamma \geq \mathrm{d}(\alpha) + \mathrm{d}(\beta) - \mathrm{d}(\gamma) \text{ and } F_A \subseteq F'.$$

But F is, by construction, the minimal ring prefiltration with this property (see 1.11), i.e., trivially $F \subseteq F'$.

7.6. Minimality of flatifying filtrations and generators of $G(A)$

Let $f : (A, m) \to (B, n)$ be a flat homomorphism of complete filtered local rings, (b, d) a distinguished basis for f, and

$$(a_{\alpha\beta}^\gamma)_{\alpha,\beta,\gamma \in \Gamma}$$

the corresponding family of structure constants. Assume that F_B is F_A-minimal with respect to (b, d). Then

(i) $G_{F(f)}(A)$ is generated as an algebra over $G_{F_A}(A)$ by the residue classes

$$(a_{\alpha\beta}^\gamma \bmod F(f)^{\mathrm{d}(\alpha) + \mathrm{d}(\beta) - \mathrm{d}(\gamma) + 1})$$

$(\alpha, \beta, \gamma \in \Gamma)$ of the structure constants.

(ii) A ring filtration F of A containing F_A such that (i) is satisfied with $F(f)$ replaced by F, coincides with the minimal flatifying filtration: $F = F(f)$.

Note that assertion (i) is a statement about a prefiltration whereas in (ii), F is assumed to be a filtration, i.e, to have the Artin-Rees property (cf 3.11).

Proof. Define

$$\left. \begin{array}{rcl} a(\omega) & := & a_{\alpha\beta}^\gamma \\ \mathrm{d}(\omega) & := & \mathrm{d}(\alpha) + \mathrm{d}(\beta) - \mathrm{d}(\gamma) \end{array} \right\} \omega := (\alpha, \beta, \gamma) \in \Omega := \Lambda \times \Lambda \times \Lambda$$

Then, by 7.5, $F(f)$ is the ring prefiltration generated over F_A by the family $(a(\omega))_{\omega \in \Omega}$ such that

$$\mathrm{ord}\, a(\omega) \geq \mathrm{d}(\omega).$$

In other words (see 1.11),

$$(1) \qquad F(f)^d := \sum_{(i,\omega)\in\mathbb{N}^n\times\Omega^n} F_A^{d-<i,d(\omega)>} \cdot a(\omega)^i$$

The sum is taken over all pairs (i,ω) where i and ω have the same number of coordinates $n(=1,2,3,\dots)$, and we are using multi-index notation:

$$< i, d(\omega) >:= i_1\, d(\omega_1) + \dots + i_n\, d(\omega_n)$$
$$a(\omega)^i := a(\omega_1)^{i_1} \cdot \dots \cdot a(\omega_n)^{i_n}$$

if $i = (i_1,\dots,i_n)$ and $\omega = (\omega_1,\dots,\omega_n)$. Since $\mathrm{ord}_{F(f)}\, a(\omega) \geq d(\omega)$ for $\omega \in \Omega$, there are well-defined elements

$$A(\omega) := (a(\omega) \bmod F(f)^{d(\omega)+1}) \in G_{F(f)}(A).$$

Moreover,

$$(2) \qquad G_{F(f)}(A)(d) = \sum_{(i,\omega)\in\mathbb{N}^n\times\Omega^n} G_{F_A}(d- < i, d(\omega) >) \cdot A(\omega)^i,$$

i.e., $G_{F(f)}(A)$ is generated as an algebra over $G_{F_A}(A)$ by the elements $A(\omega)$, $\omega \in \Omega$.

Conversely, assume that assertion (i) is true for some ring filtration F of A containing F_A. Then identity (2) holds for every d with $F(f)$ replaced by F, hence in view of (1),

$$G_F(A)(d) = \sum_{(i,\omega)\in\mathbb{N}^n\times\Omega^n} F_A^{d-<i,d(\omega)>} \cdot a(\omega)^i \bmod F^{d+1} = F(f)^d \bmod F^{d+1}.$$

Therefore $F^d = F(f)^d + F^{d+1}$ for every d. Iterating this identity, we get

$$F^d = \bigcap_{k \in \mathbb{N}} (F(f)^d + F^k).$$

Since F is assumed to be an Artin-Rees filtration, Krull's intersection theorem ([Ma86], Th. 8.10) implies $F = F(f)$.

7.7. $F(f)$ under surjective base change

Let $f : (A,m) \to (B,n)$ be a flat homomorphism of filtered local rings, and $I \subseteq A$ a proper ideal. Then the minimal flatifying prefiltrations satisfy

$$F(f \otimes_A A/I) = F(f) \cdot A/I.$$

Proof. Let

$$\alpha : A \to \bar{A} := A/I \text{ and } \beta : B \to \bar{B} := B/IB$$

be the natural homomorphisms and write

$$\bar{f} := f \otimes_A A/I : \bar{A} \to \bar{B}$$

for the homomorphism obtained from f by base change. Consider the prefiltration F such that

$$F^d := \alpha^{-1}(F(\bar{f})^d)$$

Then there are natural graded isomorphisms

(1) $\qquad \mathrm{G}_F(A) \cong \mathrm{G}_{F(\bar{f})}(\bar{A})$ and $\mathrm{G}_{F_B + f_* F}(B) \cong \mathrm{G}_{F_{\bar{B}} \cdot \bar{B} + \bar{f}_* F(\bar{f})}(\bar{B})$

The first isomorphism (1) results directly from the definition of $F := (F^d)_{d \in \mathbb{N}}$. As for the second, note that

$$(F_B + f_* F)^d = \sum_{i+j=d} F^i \cdot F_B^j + IB$$

$$= \beta^{-1}(\sum_{i+j=d} F(\bar{f})^i \cdot F_B^j \cdot \bar{B})$$

$$= \beta^{-1}((F_B \cdot \bar{B} + \bar{f}_* F(\bar{f}))^d).$$

From (1) we see that F is flatifying for f, hence $F(f) \subseteq F$. Therefore,

$$F(f) \cdot \bar{A} \subseteq F \cdot \bar{A} = F(\bar{f}).$$

In order to prove the converse inclusion, note that tangential flatness of the homomorphism $(A, F(f)) \to (B, F_B + f_* F(f))$ implies that the corresponding homomorphism induced on the residue classes modulo I,

$$(\bar{A}, F(f) \cdot \bar{A}) \to (\bar{B}, F_B \cdot \bar{B} + f_*(F(f) \cdot \bar{A}))$$

also is tangentially flat by 3.8. So $F(f) \cdot \bar{A}$ is a flatifying prefiltration for $\bar{f}$, i.e.,

$$F(\bar{f}) \subseteq F(f) \cdot \bar{A}$$

as required.

Remark

Our next aim is to show that the minimal flatifying prefiltration is even a filtration. The proof is based on the next theorem below which states that a flat homomorphism of filtered local rings becomes tangentially flat after a base change with a flat finite covering of a very special type (in particular the base change homomorphism is a complete intersection). We cannot prove the assertion in full generality but need a Noetherian assumption which is, however, satisfied for the natural filtrations.

7.8. Tangential flatification via finite base change

Let $f : (A, m) \to (B, n)$ be a homomorphism of filtered local rings such that the following conditions are satisfied.

1. f is flat.

2. $\operatorname{in}(mB)/\operatorname{in}(m) G(B)$ is a finitely generated $G(B)$-module.

Moreover let $\{\mu_1, \ldots, \mu_N\}$ be a generating system for m and k a positive integer. Consider the commutative diagram of filtered local rings

$$
\begin{array}{ccc}
(A, m) & \xrightarrow{f} & (B, n) \\
\alpha \downarrow & & \downarrow \beta \\
(A', m') & \xrightarrow{f'} & (B', n')
\end{array}
$$

where (cf 6.6)

$$A' := A[[X_1, \ldots, X_N \,|\, \operatorname{ord} X_i = 1]]/(X_1^k - \mu_1, \ldots, X_N^k - \mu_N),$$
$$B' := B[[X_1, \ldots, X_N \,|\, \operatorname{ord} X_i = 1]]/(X_1^k - \mu_1, \ldots, X_N^k - \mu_N),$$

the vertical homomorphisms are the canonical ones, and $f' = f \otimes_A A'$. Then, for k sufficiently large, $f' : A' \to B'$ is a tangentially flat homomorphism of filtered local rings.

We will see below that it is sufficient to choose k strictly greater than the degrees of any homogeneous generating system of $\operatorname{in}(mB)/\operatorname{in}(m) G(B)$.

Proof. The homomorphism $f' := f \otimes_A A'$ is flat. Since F_B is an Artin-Rees filtration, there is a function $e : \mathbb{N} \to \mathbb{N}$ such that $F_B^{e(d)} \subseteq (F_B^1)^d$. We may assume that $e(d) \geq d$ for every d. But then, if x_i denotes the residue class in B' of X_i and (x) is the ideal generated by the elements x_i,

$$F_{B'}^{2e(d)} = \sum_{i+j=2e(d)} F_B^i \cdot (x)^j B' \subseteq F_B^{e(d)} B' + (x)^{e(d)} B' \subseteq (F_{B'}^1)^d.$$

This shows that $F_{B'}$ is an Artin-Rees filtration (see 1.14(iii)). Similarly one deduces that $F_{A'}$ is an Artin-Rees filtration. The homomorphism $f' \otimes_{A'} A'/m'$ is trivially tangentially flat. To prove the claim, it will be sufficient to show (see 4.4) that $\text{in}(m'B') = \text{in}(m')\,G(B')$ for k sufficiently large. Consider the commutative diagram with exact columns,

$$
\begin{array}{ccc}
0 & & 0 \\
\downarrow & & \downarrow \\
\text{Ker}(\gamma) & \rightarrow & \text{Ker}(\gamma') \\
\downarrow & & \downarrow \\
G(B)/\text{in}(m)\,G(B) & \rightarrow & G(B')/\text{in}(m')\,G(B') \\
\gamma\downarrow & & \downarrow\gamma' \\
G(B/mB) & \rightarrow & G(B'/m'B') \\
\downarrow & & \downarrow \\
0 & & 0
\end{array}
$$

where γ and γ' are the canonical surjections and the horizontal maps are induced by β. We want to prove that

$$\text{Ker}(\gamma') = \text{in}(m'B')/\,\text{in}(m')\,G(B')$$

is zero for large k. The lower horizontal homomorphism of the diagram is an isomorphism, since

$$B'/m'B' \cong B' \otimes_{A'} A'/m' \cong B \otimes_A A'/m' \cong B/mB \otimes_{A/m} A'/m' \cong B/mB$$

(note that $\alpha : A \to A'$ is residually rational). The ring $G(B')$ is a factor ring of a polynomial ring over $G(B)$ (see 6.6),

$$G(B') = G(B)[\text{in}(x_1),\ldots,\text{in}(x_N)],$$

where $\text{in}(x_i)$ denotes the initial form in $G(B')$ of the residue class $x_i \in B'$ of X_i. Since the elements $\text{in}(x_i)$ are in $\text{in}(m')$, the horizontal homomorphism in the middle is surjective. By the snake lemma, the upper horizontal homomorphism is surjective, too. So it will be sufficient to show that the homomorphism induced by β,

$$\text{in}(mB)/\,\text{in}(m)\,G(B) \to \text{in}(m'B')/\,\text{in}(m')\,G(B')$$

is identically zero for large k. Let $m_1,\ldots,m_r \in mB$ be finitely many elements such that the initial forms of the m_i's generate $\text{in}(mB)$ modulo $\text{in}(m)\,G(B)$. By assumption 2, such elements exist. In will be sufficient to show that the residue class of $\text{in}(m_i)$ in $\text{in}(mB)/\,\text{in}(m)\,G(B)$ is mapped to zero for every i. This will be certainly the case if the order of $m_i' := \beta(m_i)$ in B' is strictly greater than the order of m_i in B,

(1) $$\text{ord}_{B'}\, m_i' > \text{ord}_B\, m_i.$$

Let

$$k > \max\{\text{ord}_B(m_i)\,|\,i = 1,\ldots,r\}.$$

Write $m_i = \sum_{j=1}^{N} b_{ij}\mu_j$ with $b_{ij} \in B$. Then $m_i' = \sum_{j=1}^{N} \beta(b_{ij})\beta(\mu_j)$. In order to prove (1), it will be sufficient to show that the elements m_i' have orders at least k, and for this in turn it suffices to prove

$$\operatorname{ord}_{B'} \beta(\mu_j) \geq k.$$

Now by construction, $\beta(\mu_j)$ is the k-th power of x_i and x_i has order at least one. So the latter inequalities are trivially satisfied.

7.9. Existence of the minimal flatifying filtration

Let $f : (A, m) \to (B, n)$ be a homomorphism of filtered local rings such that

1. f is flat.

2. $\operatorname{in}(mB)/\operatorname{in}(m)\,G(B)$ is finitely generated over $G(B)$ in degrees $< k$.

Moreover, let $\{\mu_1, \ldots, \mu_N\}$ be a generating system of the maximal ideal m and $F = (F^d)_{d \in \mathbf{N}}$ the filtration defined by

$$F^d := \sum_{i+k \cdot i_1 + \ldots + k \cdot i_N \geq d} F_A^i \cdot \mu_1^{i_1} \cdots \cdots \mu_N^{i_N}$$

Then F is a flatifying filtration for f. In particular, the minimal flatifying prefiltration $F(f)$ for f is a filtration (i.e., it has the Artin-Rees property).

Proof. Consider the commutative diagram of 7.8,

$$
\begin{array}{ccc}
(A, m) & \xrightarrow{f} & (B, n) \\
\alpha \downarrow & & \downarrow \beta \\
(A', m') & \xrightarrow{f'} & (B', n')
\end{array}
$$

with

$$A' := A[[X_1, \ldots, X_N \,|\, \operatorname{ord} X_i = 1]]/(X_1^k - \mu_1, \ldots, X_N^k - \mu_N),$$
$$B' := B[[X_1, \ldots, X_N \,|\, \operatorname{ord} X_i = 1]]/(X_1^k - \mu_1, \ldots, X_N^k - \mu_N),$$

The choice of k implies that f' is tangentially flat (see 7.8). If x_i denotes the residue class in A' of the indeterminate X_i, the filtrations of A' and B' are such that

$$F_{A'}^d := \sum_{i+i_1 + \ldots + i_N \geq d} F_A^i \cdot x_1^{i_1} \cdots \cdots x_N^{i_N}$$

$$F_{B'}^d := \sum_{i+i_1 + \ldots + i_N \geq d} F_B^i \cdot x_1^{i_1} \cdots \cdots x_N^{i_N}$$

Let F be the filtration as in the statement above. Then $\alpha : (A, F) \rightarrow (A', F_{A'})$ is a homomorphism of filtered local rings. So one has a diagram of filtered local rings,

$$
\begin{array}{ccc}
(A, F) & \overset{f}{\rightarrow} & (B, F_B + f_* F) \\
\alpha \downarrow & & \downarrow \beta \\
(A', F_{A'}) & \overset{f'}{\rightarrow} & (B', F_{B'})
\end{array}
$$

To prove that $f : (A, F) \rightarrow (B, F_B + f_* F)$ is tangentially flat, it will be sufficient to show that the following conditions are satisfied (see 6.16).

(1) $\qquad\qquad \mathrm{H}^0_\alpha = \mathrm{H}^0_\beta$

(2) $\qquad\qquad \mathrm{H}^0_f = \mathrm{H}^0_{f'}$

(3) $\qquad\qquad \alpha : (A, F) \rightarrow (A', F_{A'})$ is tangentially flat.

As for Condition (1) note that H^0_α and H^0_β are the Hilbert series of the rings (equipped with the natural filtrations),

$$A'/mA' \cong (A/m)[[X_1, \ldots, X_N]]/(X_1^k, \ldots, X_N^k)$$
$$B'/nB' \cong (B/n)[[X_1, \ldots, X_N]]/(X_1^k, \ldots, X_N^k)$$

hence are both equal to $((1 - T^k)/(1 - T))^N$ (see 6.9). So (1) is satisfied. As for condition (2) note that

$$B'/m'B' \cong B'/(m, X_1, \ldots, X_N)B' \cong B/mB.$$

Moreover,

$$F^d_{B'} := \sum_{i + i_1 + \ldots + i_N \geq d} F^i_B \cdot X_1^{i_1} \cdot \ldots \cdot X_N^{i_N} \cdot B'$$

and the filtration induced by $F_{B'}$ on $B'/m'B'$ corresponds under this isomorphism to the filtration induced by F_B on B/mB. So condition (2) is also satisfied, and the proof will be complete (by 6.13) if we can show that

$$\mathrm{H}^1_{A'} = \mathrm{H}^1_{(A,F)} \cdot \mathrm{H}^0_\alpha .$$

The element μ_i has order $\geq k$ in (A, F). So $f_i := X_i^k - \mu_i$ is a power series of order k in $R := (A, F)[[X_1, \ldots, X_N \mid \mathrm{ord}\, X_i = 1]]$ and one has $A' \cong R/(f_1, \ldots, f_N)$ as filtered local rings. Therefore by 6.9,

$$\mathrm{H}^N_{A'} \geq \mathrm{H}^N_{(A,F)} \cdot \left(\frac{1 - T^k}{1 - T} \right)^N = \mathrm{H}^N_{(A,F)} \cdot \mathrm{H}^0_\alpha .$$

The converse inequality follows from 6.10, hence (3) is true. Note that $F = F_A + F'$ where F' is the $(k, \ldots, k)$-adic filtration defined by $(\mu_1, \ldots, \mu_N)$ (see 1.5). In particular, F is an Artin-Rees filtration (by 1.15).

8. Kodaira-Spencer maps

Remark

Below we will define the Kodaira-Spencer map associated with a flat homomorphism $f : (A, m) \to (B, n)$ of filtered local rings, where the filtrations are such that f is tangentially flat. In fact the Kodaira-Spencer map will only depend upon the homomorphism $G(A) \to G(B)$ induced on the associated graded rings so that most constructions will be carried out in the context of graded rings. The notion of 'permissible graded homomorphism' which we are going to introduce now, summarizes the properties of $G(A) \to G(B)$ that will be needed. In fact the constructions are possible in a much more general situation. The case given here is just the one we need.

8.1. Permissible graded algebras and homomorphisms

A graded ring G is called *normally flat*, if it is flat as a module over its degree zero part $G(0)$. Let (A_0, m_0) be an Artinian local ring. A *permissible graded algebra* over A_0 is a graded ring G with finite Hilbert function and such that $G(0) = A_0$. Let $f_0 : (A_0, m_0) \to (B_0, n_0)$ be a homomorphism of Artinian local rings (which is automatically local). A *permissible graded homomorphism* over f_0 is a graded ring homomorphism $f : G \to G'$ with the following properties.

(i) $f : G \to G'$ is flat.

(ii) f_0 is the degree zero part of f (in particular, $A_0 = G(0)$ and $B_0 = G'(0)$).

(iii) G and G' have finite Hilbert functions.

(iv) $G'/(m_0, G^+)G'$ is normally flat.

An important consequence of these conditions that will be frequently used is that the homomorphism

$$f_{B_0} : G_{B_0} := G \otimes_{A_0} B_0 \to G', g \otimes b \mapsto f(g) \cdot b$$

induced by f is also flat.

In order to prove this, let $I := m_0 G + G^+$ and equip the rings G, G_{B_0}, and G' with the I-adic filtrations. Then, by the local flatness criterion ([Ma86], Th. 22.3(4')), the canonical surjections

$$G' \otimes_G G_I(G) \to G_I(G') \text{ and } G_{B_0} \otimes_G G_I(G) \to G_I(G_{B_0})$$

are bijective. Note that G_{B_0} is flat over G (since B_0 is A_0-flat by (i)). So there are canonical isomorphisms

$$G' \otimes_{G_{B_0}} G_I(G_{B_0}) \cong G' \otimes_{G_{B_0}} G_{B_0} \otimes_G G_I(G) \cong G' \otimes_G G_I(G),$$

i.e., the canonical surjection $G' \otimes_{G_{B_0}} G_I(G_{B_0}) \to G_I(G')$ is also bijective. We cannot directly use now the local flatness criterion as formulated in [Ma86] to conclude that G' is flat over G_{B_0}, since we do not assume that G is Noetherian. But at any case the criterion implies that $G'/I^d G'$ is flat over $G_{B_0}/I^d G_{B_0}$ for every non-negative integer d. Note that the case $d = 1$ is just condition (iv) above, for, $G_{B_0}/IG_{B_0} \cong (G/I) \otimes_{A_0} B_0 \cong A_0/m_0 \otimes_{A_0} B_0 \cong B_0/m_0 B_0 \cong (G'/IG')(0)$. Since the degree 0 part of I is nilpotent, the graded variant 2.3 of the local flatness criterion implies that G' is flat over G_{B_0} (cf the third remark of 3.11).

Example

Let $f : (A, m) \to (B, n)$ be a tangentially flat homomorphism of filtered local rings. Then the induced homomorphism $G(f) : G(A) \to G(B)$ satisfies the conditions on a permissible graded homomorphism over $G(f)(0) : A/F_A^1 \to B/F_B^1$ except for, possibly, condition (iv). In case

$$F_B^1 = n$$

this latter condition is satisfied, too.

8.2. Generalized graded structures

Let G be a graded ring. Below there will be constructed modules over G having a graded structure in the following slightly weaker sense. A G-module M will be called *weakly graded*, if it is equipped with a family of $G(0)$-submodules $M(d) \subseteq M$, $d \in \mathbb{Z}$, such that

$$\bigoplus_{d \in \mathbb{Z}} M(d) \subseteq M \subseteq \prod_{d \in \mathbb{Z}} M(d)$$

and $G(d) \cdot M(d') \subseteq M(d+d')$ for arbitrary $d \in \mathbb{N}$, $d' \in \mathbb{Z}$. As in the case of usual gradings (cf 1.1) we assume that $M(d) = 0$ for all d less than a fixed value.

8.3. Schlessinger's T^1 of a graded ring

Let (A_0, m_0) be an Artinian local ring and G a permissible graded algebra over A_0. Choose a family

$$g = (g_\gamma)_{\gamma \in \Gamma}$$

of homogeneous elements from G^+ that generate G as an algebra over $G(0) = A_0$. Further, choose a family

$$X = (X_\gamma)_{\gamma \in \Gamma}$$

of indeterminates and consider the polynomial ring $A_0[X]$ as a graded ring such that $\deg X_\gamma := \deg g_\gamma$ and such that the elements of A_0 have degree zero. Then there is a surjective homomorphism of graded rings

$$p_g : A_0[X] \to G, X_\gamma \mapsto g_\gamma, p_g|_{A_0} = id_{A_0}.$$

We will use the following notation.

$$I_g := \mathrm{Ker}(p_g), \qquad N_g := \mathrm{Hom}_{A_0[X]}(I_g, G), \qquad D_g := \mathrm{Der}_{A_0}(A_0[X], G).$$

The ideal I_g is called the *defining ideal* of G and N_g the *normal module* of G (with respect to the embedding given by the family g). D_g is the G-module of *derivations* $A_0[X] \to G$ over A_0. Note that the product formula for derivations implies

$$\alpha P(X) = \sum_{\gamma \in \Gamma} \frac{\partial P(X)}{\partial X_\gamma} \cdot \alpha X_\gamma \text{ for } P \in A_0[X], \alpha \in D_g.$$

The modules N_g and D_g have a weakly graded structure with

$$N_g(d) = \{\alpha \in N_g | \alpha(I_g(n)) \subseteq G(n + d) \text{ for every } n \in \mathbb{N}\},$$
$$D_g(d) = \{\alpha \in D_g | \alpha(A_0[X](n)) \subseteq G(n + d) \text{ for every } n \in \mathbb{N}\}$$

(with $I_g(d) := I_g \cap A_0[X](d)$). Moreover, there is a natural G-linear map

$$r_g : D_g \to N_g, \alpha \mapsto \alpha|_{I_g}$$

respecting the weakly graded structures. To see that it is well-defined let $\alpha \in D_g$, $x \in I_g$, $y \in A_0[X]$. Then, since I_g annihilates G,

$$\alpha(y \cdot x) = y \cdot \alpha(x) + x \cdot \alpha(y) = y \cdot \alpha(x),$$

i.e., $\alpha|_{I_g}$ is $A_0[X]$-linear. The cokernel of this map,

$$T_G^1 := T_g^1 := \mathrm{Coker}(r_g : D_g \to N_g)$$

is called *Schlessinger's* T^1 of the permissible graded algebra G. The statement that follows says that T_G^1 is independent (up to isomorphism) upon the special choice of the generating system g.

8.4. Invariance of the definition

Let (A_0, m_0) be an Artinian local ring, and G a permissible graded ring over A_0. Further let

$$g = (g_\gamma)_{\gamma \in \Gamma} \text{ and } g' = (g'_{\gamma'})_{\gamma' \in \Gamma'}$$

be two homogeneous generating systems for G over A_0. Then the modules T_g^1 and $\mathrm{T}_{g'}^1$ are canonically isomorphic.

Proof. Comparing the families g, g', and $g \cup g'$, the claim is easily reduced to the case that g' is a subfamily of g, i.e., $\Gamma' \subseteq \Gamma$ and $g'_{\gamma'} = g_{\gamma'}$ for $\gamma' \in \Gamma'$. In order to prove the assertion for this case, let $X = (X_\gamma)_{\gamma \in \Gamma}$ be a family of indeterminates and $X' = (X_\gamma)_{\gamma \in \Gamma'}$ the subfamily corresponding to the generating system g'. Consider the commutative diagram with exact rows

$$
\begin{array}{ccccccccc}
0 & \to & I_g & \to & A_0[X] & \overset{p_g}{\to} & G & \to & 0 \\
& & \uparrow & & \uparrow & & \| & & \\
0 & \to & I_{g'} & \to & A_0[X'] & \overset{p_{g'}}{\to} & G & \to & 0
\end{array}
$$

where the vertical homomorphism in the middle is the natural inclusion. Identify $A_0[X']$ with a subring of $A_0[X]$ via this map. Then I_g is generated by the elements of $I_{g'}$ together with certain linear polynomials of type

$$
X_\gamma - P_\gamma(X') \qquad \text{with } P_\gamma(X') \in A_0[X'], \quad \gamma \in \Gamma - \Gamma'.
$$

We may assume that $P_\gamma = 0$, so that I_g is generated by $I_{g'}$ and the variables of the difference set $X - X'$,

$$
I_g = (I_{g'}, X - X')A_0[X'].
$$

Consider the commutative diagram with exact rows,

$$
\begin{array}{ccccccccc}
0 & \to & \mathrm{Ker}(\delta) & \to & D_g & \overset{\delta}{\to} & D_{g'} & \to & 0 \\
& & \mu \downarrow & & \downarrow r_g & & \downarrow r_{g'} & & \\
0 & \to & \mathrm{Ker}(\nu) & \to & N_g & \overset{\nu}{\to} & N_{g'} & \to & 0
\end{array}
$$

where the horizontal maps ν and δ are induced by the natural inclusions $I_{g'} \subseteq I_g$ and $A_0[X'] \subseteq A_0[X]$, respectively, and μ exists due to the fact that the square on the right is commutative. The maps δ and ν are surjective since the inclusion $A_0[X'] \to A_0[X]$ has a left inverse $A_0[X] \to A_0[X']$ (mapping the variables of $X - X'$ to zero) which maps I_g onto $I_{g'}$. Taking cokernels, we get a homomorphism

$$
\mathrm{T}_g^1 \to \mathrm{T}_{g'}^1
$$

which is surjective since ν is surjective. We want to show that it is also injective. By the snake lemma, it will be sufficient to show that μ is surjective. Let $\alpha \in N_g$ be in the kernel of ν. Define a derivation $d \in D_g$ such that

$$
dP(X) := \sum_{\gamma \in \Gamma - \Gamma'} \frac{\partial P(X)}{X_\gamma} \cdot \alpha X_\gamma.
$$

Note that d is well-defined since the variables $X_\gamma \in X - X'$ are in the ideal $I_g = (I_{g'}, X - X')A_0[X]$ and α is defined on I_g. Obviously $d|_{A_0[X']} = 0$, i.e., d is in the kernel of δ. It will be sufficient to show that d is mapped to α, i.e.,

$$d|_{I_g} = \alpha.$$

A polynomial $P(X) \in I_g$ can be written

$$(1) \quad P(X) = Q(X') + \sum_{\gamma \in \Gamma - \Gamma'} P_\gamma(X) \cdot X_\gamma \quad \text{with } Q \in I_{g'}, \quad P_\gamma \in A_0[X].$$

Since α is in the kernel of ν (i.e., $\alpha|_{I_{g'}} = 0$), and since α is an $A_0[X]$-linear mapping,

$$\alpha P(X) = \sum_{\gamma \in \Gamma - \Gamma'} P_\gamma(X) \cdot \alpha X_\gamma.$$

This is almost the definition of d. We have yet to show that the coefficients $P_\gamma(X)$ can be replaced by the partial derivatives of $P(X)$. Differentiating (1) we see that for $\gamma \in \Gamma - \Gamma'$,

$$\frac{\partial P(X)}{\partial X_\gamma} \equiv P_\gamma(X) \bmod (X - X').$$

Since G is annihilated by the variables from $X - X'$ ($\subseteq I_g$), we deduce $\alpha P = dP$ for $P \in I_g$, i.e., $d|_{I_g} = \alpha$ as required.

8.5. Embedded Kodaira-Spencer map

Let $f_0 : (A_0, m_0) \to (B_0, n_0)$ be a flat local homomorphism of Artinian local rings and $f : G \to G'$ a permissible graded map over f_0. Define

$$\bar{B}_0 := B_0/m_0 B_0, \quad \bar{G} := G'/m'G', \quad m' := (m_0, G^+)G.$$

Fix surjective graded algebra homomorphisms

$$p : A_0[X] \to G \quad \text{and} \quad \bar{p} : \bar{B}_0[Y] \to \bar{G}$$

which restrict to the identity maps on A_0 and B_0, respectively, and where $X = (X_\gamma)_{\gamma \in \Gamma}$ and $Y := (Y_\nu)_{\nu \in N}$ are families of indeterminates with appropriate non-negative degrees. In that what follows the X_γ's will be assumed to have even positive degrees,

$$\deg(X_\gamma) > 0 \quad \text{for every } \gamma.$$

Of course, we could equally assume this for the indeterminates in Y. But later it will turn out to be useful if we allow more flexibility here. Next choose a homogeneous representative $g'_\nu \in G'$ for every $\bar{p}(Y_\nu) \in \bar{G}$ and consider the homomorphism $p' : B_0[X, Y] \to G'$ of graded B_0-algebras mapping X_γ to $f(p(X_\gamma)) \in G'$

and Y_ν to $g'_\nu \in G'$. This homomorphism fits into a commutative diagram of graded algebra homomorphisms,

$$
(1) \qquad
\begin{array}{ccccc}
G & \xrightarrow{f} & G' & \xrightarrow{c} & \bar{G} \\
\uparrow p & & \uparrow p' & & \uparrow \bar{p} \\
A_0[X] & \xrightarrow{\phi} & B_0[X,Y] & \xrightarrow{\chi} & \bar{B}_0[Y]
\end{array}
$$

We will refer below to such a diagram as to an *embedding of the permissible graded homomorphism* $f : G \to G'$. Here c and χ are the canonical surjections and ϕ is the A_0-algebra homomorphism mapping each X_γ into itself (and restricting to f_0 on A_0). Note that p and $\bar{p}$ are surjective by construction and that p' is surjective since p and $\bar{p}$ are. The latter follows from the graded Nakayama lemma 2.1 in view of the identity,

$$
G' = \mathrm{Im}(p') + (m, X)G'.
$$

Note that, since $\bar{p}$ is surjective, there exists for every $x \in G'$ some element $h \in B_0[X,Y]$ with $x - p'(h) \in \mathrm{Ker}(G' \to \bar{G}) = (m, G^+)G'$, i.e., x is as required in $\mathrm{Im}(p') + (m, X)G'$. Define

$$
I := \mathrm{Ker}(p), \quad I' := \mathrm{Ker}(p'), \quad \bar{I} := \mathrm{Ker}(\bar{p}).
$$

The *embedded Kodaira-Spencer map* associated with the embedding (1) of the permissible graded homomorphism $f : G \to G'$ is defined to be the $\bar{B}_0$-linear map

$$
D : T_{G,\bar{B}_0} := \mathrm{Hom}_{A_0}(G^+/(G^+)^2, \bar{B}_0) \to \mathrm{Hom}_{\bar{B}_0}(\bar{I}, \bar{G}), \quad l \mapsto D_l,
$$

with

$$
D_l(\bar{P}(Y)) := \bar{p}\left(\sum_{\gamma \in \Gamma} \frac{\partial P'(0,Y)}{\partial X_\gamma} \cdot c_\gamma \right),
$$
$$
c_\gamma := l_*(X_\gamma) := l(p(X_\gamma) \bmod (G^+)^2)
$$

Here $P'(X,Y) \in I'$ denotes any lift of the polynomial $\bar{P}(Y) \in \bar{I}$. Geometrically, we take an equation P' of the total space $Spec(G')$ that restricts on the special fibre to the given equation $\bar{P}$ and form the differential quotient in the direction given by the tangent vector l to the parameter space $Spec(G)$. This just describes infinitesimally the movement of a point on the varying fibre of $Spec(G') \to Spec(G)$ over a point in the parameter space which is moved into direction l.

We have to show the embedded Kodaira-Spencer mapping is well-defined. First note that there is always a lift $P(X,Y) \in I'$ of $\bar{P}(Y) \in \bar{I}$ since the ideal I' maps surjectively onto $\bar{I}$. The latter is easily seen from the following commutative diagram with exact rows and columns using the nine lemma.

$$
\begin{array}{ccccccccc}
& & 0 & & 0 & & 0 & & \\
& & \uparrow & & \uparrow & & \uparrow & & \\
0 & \to & (m, G^+) & \to & G' & \to & \bar{G} & \to & 0 \\
& & \uparrow & & \uparrow p' & & \uparrow \bar{p} & & \\
0 & \to & (m, X) & \to & B_0[X, Y] & \xrightarrow{\chi} & \bar{B}_0[Y] & \to & 0 \\
& & \uparrow & & \uparrow & & \uparrow & & \\
& & I' \cap (m, X) & \to & I' & \to & \bar{I} & & \\
& & \uparrow & & \uparrow & & \uparrow & & \\
& & 0 & & 0 & & 0 & &
\end{array}
$$

Let $l \in \mathrm{Hom}_{A_0}(G^+/(G^+)^2, \bar{B}_0)$ be some element. We have to show that D_l is a well defined element of $\mathrm{Hom}_{\bar{B}_0}(\bar{I}, \bar{G})$. At any case there is a derivation $D'_l : B_0[X, Y] \to \bar{B}_0[Y] \to \bar{G}$ with

$$
D'_l(P'(X, Y)) := \bar{p}\left(\sum_{\gamma \in \Gamma} \frac{\partial P'(0, Y)}{\partial X_\gamma} \cdot c_\gamma\right)
$$

and $c_\gamma := l_*(X_\gamma)$ as above. Here the product of the polynomial $\frac{\partial P'(0, Y)}{\partial X_\gamma} \in B_0[Y]$ with the element $c_\gamma \in \bar{B}_0 \subseteq \bar{B}_0[Y]$ is defined via the canonical surjection $B_0[Y] \to \bar{B}_0[Y]$. Note that the restriction of D'_l to I' is $B_0[X, Y]$-linear by the product rule for derivations and in view of the fact that I' annihilates the image ring $\bar{G}$,

$$
D'_l(F' \cdot P') = F' \cdot D'_l(P') + P' \cdot D'_l(F') = F' \cdot D'_l(P')
$$

for $F' \in B_0[X, Y]$, $P' \in I'$. Thus all we have to show is that the value of D'_l doesn't change when $P'(X, Y)$ is modified by an element which maps to zero in $\bar{I}$. Since D'_l is additive, it will be sufficient to show that $D'_l(P'(X, Y)) = 0$ if $P'(X, Y) \in I'$ represents the zero element in $\bar{I}$. Our assumption means that $P'(X, Y) \in I' \cap (m, X)$. Define $G^* := B_0[X, Y]/(\phi(I))$ and consider the exact sequence of graded G-modules

$$
0 \to I'G^* \to G^* \to G' \to 0.
$$

Since G' is flat over G, the homomorphism

$$
(I'G^* \to G^*) \otimes_G G/(m, G^+) = I'/((m, X)I' + (\phi(I))) \to B_0[X, Y]/(m, X)
$$

is injective, i.e.,

$$
P'(X, Y) \in I' \cap (m, X) = (m, X)I' + (\phi(I)).
$$

Since the derivation D'_l takes values in the ring $\bar{G}$, which is annihilated by I' and by (m, X), it is identically zero on the product $(m, X)I'$, i.e., we have only to show $D'_l(\phi(I)) = 0$. Since f_0 is flat hence injective, we can identify A_0 with a subring of B_0 and $A_0[X]$ with a subring of $B_0[X]$. Let $P(X) \in I$. Then, by definition,

$$D'_l(P) := \bar{p}(\sum_{\gamma \in \Gamma} \frac{\partial P(0)}{\partial X_\gamma} \cdot l_*(X_\gamma)).$$

Extend l_* to an A_0-linear map

$$l_* : (X)A_0[X] \xrightarrow{p} G^+ \xrightarrow{r} G^+/(G^+)^2 \xrightarrow{l} \bar{B}_0.$$

where r is induced by the natural homomorphism $G \twoheadrightarrow G/(G^+)^2$. The polynomial $P(X) \in I$ has a free member equal to zero, for, p induces the identity map on A_0, i.e., $I \subseteq (X)A_0[X]$. Since the non-linear terms of $P(X)$ are mapped to zero under l_*, P and its linear part have the same image. Therefore, since $P \in I = \text{Ker}(p)$,

$$\sum_{\gamma \in \Gamma} \frac{\partial P(0)}{\partial X_\gamma} \cdot l_*(X_\gamma) = l_*(\sum_{\gamma \in \Gamma} \frac{\partial P(0)}{\partial X_\gamma} \cdot X_\gamma) = l_*(P) = 0,$$

i.e., $D'_l(P) = 0$, as required.

8.6. Kodaira-Spencer map

Let $f_0 : (A_0, m_0) \to (B_0, n_0)$ be a flat local homomorphism of Artinian local rings, $f : G \to G'$ a permissible graded map over f_0, and

$$
\begin{array}{ccccc}
G & \xrightarrow{f} & G' & \xrightarrow{c} & \bar{G} \\
p\uparrow & & p'\uparrow & & \uparrow\bar{p} \\
A_0[X] & \xrightarrow{\phi} & B_0[X,Y] & \xrightarrow{X} & \bar{B}_0[Y]
\end{array}
$$

an embedding of $f : G \to G'$ as defined in 8.5. Then the composition

$$Df : T_{G,\bar{B}_0} := \text{Hom}_{A_0}(G^+/(G^+)^2, \bar{B}_0) \xrightarrow{D} N_I := \text{Hom}_{B_0}(\bar{I}, \bar{G}) \xrightarrow{r} T^1_{\bar{G}}$$

where D is the embedded Kodaira-Spencer map for the given embedding and r is the canonical surjection (see 8.3) is called *Kodaira-Spencer map* of f.

Note that $G^+_{B_0} := G^+ \otimes_{A_0} B_0$ is just the ideal of $G_{B_0} := G \otimes_{A_0} B_0$ generated by the elements of positive degree from G. Therefore the Hom-module of A_0-linear maps on the left can be identified with a module of B_0-linear maps as follows,

$$\text{Hom}_{A_0}(G^+/(G^+)^2, \bar{B}_0) \cong \text{Hom}_{B_0}(G^+/(G^+)^2 \otimes_{A_0} B_0, \bar{B}_0)$$
$$\cong \text{Hom}_{B_0}(G^+_{B_0}/(G^+_{B_0})^2, \bar{B}_0)$$

We will frequently use this fact below. Our aim in this paragraph is to prove that Df doesn't depend upon the special choice of the embedding. In order to do this, let

$$
\begin{array}{ccccc}
G & \xrightarrow{f} & G' & \xrightarrow{c} & \bar{G} \\
{\scriptstyle p^*}\uparrow & & {\scriptstyle p^{*\prime}}\uparrow & & \uparrow{\scriptstyle \bar{p}^*} \\
A_0[X^*] & \xrightarrow{\phi^*} & B_0[X^*,Y^*] & \xrightarrow{\chi^*} & \bar{B}_0[Y^*]
\end{array}
$$

be a second embedding of f. Note that the embedding is completely determined by the indeterminates from X^* and Y^* and their images in G and G', respectively. So taking the unions $X \cup X^*$ and $Y \cup Y^*$ one obtains a third embedding which 'dominates' the two given ones. It will be sufficient to compare each of the originally given embeddings with the dominating one. So it is sufficient to prove the claim in case that $X^* \subseteq X$, $Y^* \subseteq Y$, and that p^*, $p^{*\prime}$ are restrictions of p and p', respectively. In other words, we have a commutative diagram

(1)
$$
\begin{array}{ccccc}
G & \xrightarrow{f} & G' & \xrightarrow{c} & \bar{G} \\
{\scriptstyle p}\uparrow & & {\scriptstyle p'}\uparrow & & \uparrow{\scriptstyle \bar{p}} \\
A_0[X] & \xrightarrow{\phi} & B_0[X,Y] & \xrightarrow{\chi} & \bar{B}_0[Y] \\
{\scriptstyle i}\uparrow & & {\scriptstyle i'}\uparrow & & \uparrow{\scriptstyle \bar{i}} \\
A_0[X^*] & \xrightarrow{\phi^*} & B_0[X^*,Y^*] & \xrightarrow{\chi^*} & \bar{B}_0[X^*]
\end{array}
$$

where the upper part of the diagram is the dominating one of the given embeddings, the top and the bottom rows come from the other one, and the vertical maps in the lower part are the canonical embeddings induced by the inclusions $X^* \subseteq X$ and $Y^* \subseteq Y$. This diagram defines a square

$$
\begin{array}{ccccc}
\mathrm{Hom}_{A_0}(G^+/(G^+)^2, \bar{B}_0) & \xrightarrow{D} & \mathrm{Hom}_{\bar{B}_0[Y]}(\mathrm{Ker}(\bar{p}), \bar{G}) & \to & \mathrm{T}^1_{\bar{G}} \\
\| & & \downarrow{\scriptstyle r} & & \| \\
\mathrm{Hom}_{A_0}(G^+/(G^+)^2, \bar{B}_0) & \xrightarrow{D^*} & \mathrm{Hom}_{\bar{B}_0[Y^*]}(\mathrm{Ker}(\overline{pi}), \bar{G}) & \to & \mathrm{T}^1_{\bar{G}}
\end{array}
$$

where D and D^* are the embedded Kodaira-Spencer maps, and the horizontal maps on the right are the canonical surjections. We have to show, this diagram is commutative for some suitably chosen homomorphism r. Identify $\bar{B}_0[X^*]$ with a subring of $\bar{B}_0[X]$ via $\bar{i}$. Then $\mathrm{Ker}(\bar{p} \cdot \bar{i})$ becomes a subset of $\mathrm{Ker}(\bar{p})$. Define r to be the homomorphism induced by the inclusion $\mathrm{Ker}(\bar{p} \cdot \bar{i}) \subseteq \mathrm{Ker}(\bar{p})$. Then the right hand side square of the diagram is commutative (as we have seen in the invariance proof for Schlessinger's T^1 in 8.4), i.e., it is sufficient to show that the left hand side square is also commutative. Write $X^* = (X_\gamma)_{\gamma \in \Gamma^*}$ and $X = (X_\gamma)_{\gamma \in \Gamma}$ with $\Gamma^* \subseteq \Gamma$. By definition

$$
D_l(\bar{P}) = \bar{p}\left(\sum_{\gamma \in \Gamma} \frac{\partial P'(0,Y)}{\partial X_\gamma} \cdot l_*(X_\gamma) \right)
$$

for every linear form $l : G^+/(G^+)^2 \to \bar{B}_0$, every $\bar{P} \in \mathrm{Ker}(\bar{p})$ and every lift $P' \in \mathrm{Ker}(p')$ of $\bar{P}$. Now assume $\bar{P}$ is even in $\mathrm{Ker}(\bar{p} \cdot \bar{i})$. Then the lift P' of $\bar{P}$

may be even taken from $\mathrm{Ker}(p' \cdot i')$, and it is sufficient to take the above sum over $\gamma \in \Gamma^*$,

$$D_l(\bar{P}) = \bar{p}\left(\sum_{\gamma \in \Gamma^*} \frac{\partial P'(0, Y)}{\partial X_\gamma} \cdot l_*(X_\gamma)\right).$$

But the latter expression is just $D_l^*(\bar{P})$, so that $D_l|_{\mathrm{Ker}(\bar{p} \cdot \bar{i})} = D_l^*$, hence $r \cdot D = D^*$, and the square on the left is commutative, too.

8.7. Base change via surjections

Let $f_0 : (A_0, m) \to (B_0, n)$ be a flat local homomorphism of Artinian local rings, $f : G \to G'$ a permissible graded map over f_0, and $I \subseteq G$ a proper homogeneous ideal. Then the Kodaira-Spencer maps Df and Df^* of f and $f^* := f \otimes_G G/I$, respectively, fit into a commutative diagram

$$\begin{array}{ccc}
\mathrm{Hom}_{A_0}(G^+/(G^+)^2, \bar{B}_0) & \overset{Df}{\to} & T_{\bar{G}}^1 \\
{\scriptstyle r} \uparrow & & \| \\
\mathrm{Hom}_{A_0}((I + G^+)/(I + (G^+)^2), \bar{B}_0) & \overset{Df^*}{\to} & T_{\bar{G}}^1
\end{array}$$

where the vertical map r on the left is induced by the canonical surjection $G^+/(G^+)^2 \to (I + G^+)/(I + (G^+)^2)$.

Note that r is an isomorphism if $I \subseteq mG^+ + (G^+)^2$ (since $\bar{B}_0 = B_0/mB_0$ is annihilated by m).

Proof. Let

$$\begin{array}{ccccc}
G & \overset{f}{\to} & G' & \overset{c}{\to} & \bar{G} \\
p \uparrow & & p' \uparrow & & \uparrow \bar{p} \\
A_0[X] & \overset{\phi}{\to} & B_0[X, Y] & \overset{X}{\to} & \bar{B}_0[Y]
\end{array}$$

be an embedding of f. Tensor the upper row with G/I over G and the lover row with $A_0/I(0)$ over A_0. This yields an embedding of f^*,

$$\begin{array}{ccccc}
G^* & \overset{f^*}{\to} & G'^* & \overset{c^*}{\to} & \bar{G} \\
p^* \uparrow & & p'^* \uparrow & & \uparrow \bar{p} \\
A_0^*[X] & \overset{\phi^*}{\to} & B_0^*[X, Y] & \overset{X^*}{\to} & \bar{B}_0[Y]
\end{array}$$

where

$$\begin{array}{ll}
G^* := G/I, & G'^* := G'/IG', \\
A_0^* := A_0/I(0)A_0, & B_0^* := B_0/I(0)B_0
\end{array}$$

Note that $I \subseteq mG + G^+$ since I is proper and homogeneous. Let $N_{\bar{I}}$ be the normal module $N_{\bar{I}} = \mathrm{Hom}_{\bar{B}_0}(\bar{I}, \bar{G})$ and

$$D : \mathrm{Hom}_{A_0}(G^+/(G^+)^2, \bar{B}_0) \to N_{\bar{I}}$$
$$D^* : \mathrm{Hom}_{A_0}(G^{*+}/(G^{*+})^2, \bar{B}_0) \to N_{\bar{I}},$$

the embedded Kodaira-Spencer maps associated with the above embeddings. Consider a linear form $l^* : G^{*+}/(G^{*+})^2 \to \bar{B}_0$ and denote by $l : G^+/(G^+)^2 \to \bar{B}_0$ the composition of l^* with the canonical projection $G^+/(G^+)^2 \to G^{*+}/(G^{*+})^2$. Then it will be sufficient to show that $D^*_{l^*} = D_l$.

Given an element $\bar{P}$ from the kernel of $\bar{p}$ choose some lift $P' \in \mathrm{Ker}(p')$ of $\bar{P}$ Then

$$D_l(\bar{P}) := \bar{p}\left(\sum_{\gamma \in \Gamma} \frac{\partial P'(0, Y)}{\partial X_\gamma} \cdot l(p(X_\gamma) \bmod (G^+)^2)\right)$$

Moreover, the residue class of P' in $B^*_0[X, Y]$ is a lift of $\bar{P}$ suited to calculate $D^*_{l^*}(\bar{P})$, and the formula for $D^*_{l^*}(\bar{P})$ is essentially the same as for $D_l(\bar{P})$. The only difference is that l must be replaced by l^* and that the partial derivatives are formed in $B^*_0[Y]$ instead of $B_0[Y]$ (and then are mapped to $\bar{B}_0[Y]$). Therefore $D^*_{l^*}(\bar{P}) = D^*_{l^*}(\bar{P})$, as required.

8.8. Linear maps definable over a subalgebra

Let $f_0 : (A_0, m_0) \to (B_0, n_0)$ be a flat local homomorphisms of local rings. A B_0-module *defined over A_0* is by definition a pair (M, M_{A_0}) consisting of a B_0-module M and an A_0-submodule M_{A_0} such that the canonical map $M_{A_0} \otimes_{A_0} B_0 \to M$, $m \otimes b \mapsto m \cdot b$ is an isomorphism. Note that for free modules this means that the following two conditions are satisfied

1. The rank of M_{A_0} over A_0 equals the rank of M over B_0 , $\mathrm{rank}_{A_0} M_{A_0} = \mathrm{rank}_{B_0} M$.

2. The submodule M_{A_0} generates M over B_0.

Whenever it is clear from the context which submodule M_{A_0} of M we have in mind, we will also say that M is a module defined over A_0 and refer to M_{A_0} as to the *defining submodule* of M.

Let M and M' be B_0-modules defined over A_0. A B_0-linear map $l : M \to M'$ is called to be *defined over A_0*, if it respects the defining subspaces, i.e., $l(M_{A_0}) \subseteq M'_{A_0}$ or, equivalently, if there is a A_0-linear map $l_{A_0} : M_{A_0} \to M'_{A_0}$ with $l_{A_0} \otimes_{A_0} B_0 = l$. In other words, the matrix of l in terms of generators taken from the defining submodules has all its entries in A_0. Since we assume that $f_0 : (A_0, m_0) \to (B_0, n_0)$ is flat, a B_0-linear map l defined over A_0 satisfies

$$\mathrm{Ker}(l) = \mathrm{Ker}(l_{A_0}) \otimes_{A_0} B_0 \quad \text{and} \quad \mathrm{Im}(l) = \mathrm{Im}(l_{A_0}) \otimes_{A_0} B_0.$$

In particular, l is injective (respectively surjective) if and only if so is l_{A_0}. A slight generalization of the above concept might be useful to prove generalized versions of our main result 9.2 (see Problem 9.6). Given B_0-modules M and M' such that M is defined over A_0, a B_0-linear map $l : M \to M'$ is called to be *definable over A_0* if there is an injective B_0-linear map $l' : M' \to M''$ with some B_0-module M'' defined over A_0 such that the composition $l' \circ l : M \to M''$ is defined over A_0.

Example 1

Consider the complex vector space $\mathbb{C}^2$ defined over $\mathbb{R}$ with the defining subspace equal to $\mathbb{R}^2$. Then the $\mathbb{C}$-linear map

$$l : \mathbb{C}^2 \to \mathbb{C}, \quad (x,y) \mapsto x + i \cdot y$$

is not definable over $\mathbb{R}$, for, the restriction $l|_{\mathbb{R}^2}$ of l to the defining subspace is injective whereas the map l itself is not, a phenomenon that cannot happen for a map definable over $\mathbb{R}$.

Example 2

Let $f_0 : (A_0, m_0) \to (B_0, n_0)$ be a flat local homomorphism of Artinian local rings and $f : G \to G'$ a permissible graded map over f_0. Then the homogeneous parts of the B_0-module

$$T_{G,\bar{B}_0} := \operatorname{Hom}_{A_0}(G^+/(G^+)^2, \bar{B}_0), \quad \bar{B}_0 := B_0/m_0 B_0$$

(which is graded in the generalized sense, cf 8.2) are defined over A_0 with the defining subspace of $T_{G,B_0}(d)$ equal to the degree d part of

$$T_{G,\bar{A}_0} := \operatorname{Hom}_{A_0}(G^+/(G^+)^2, \bar{A}_0), \quad \bar{A}_0 := A_0/m_0$$

In case the ring G is Artinian (hence a finitely generated A_0-module) the B_0-module $T_{G,\bar{B}_0}$ itself is defined over A_0 with the defining submodule equal to $T_{G,\bar{A}_0}$.

Note that, in the general (non-Artinian) case,

$$T_{G,\bar{A}_0}(d) \otimes_{A_0} B_0 = \operatorname{Hom}_{A_0}(G^+/(G^+)^2(-d), \bar{B}_0) = T_{G,B_0}(d)$$

This is just a consequence of the fact that G has finite Hilbert function and that the Hom-functor commutes with the tensor product by flat modules, provided the first argument is finitely presented (see [Ma86], Th. 7.11).

Below we will equip, whenever appropriate, $T_{G,\bar{B}_0}$ and $T_{G,\bar{B}_0}(d)$ with the indicated structures of modules defined over A_0.

8.9. Extending coefficients

Let $f_0 : (A_0, m_0) \to (B_0, n_0)$ be a flat local homomorphism of Artinian local rings, $f : G \to G'$ a permissible graded map over f_0, and

$$(1) \quad \begin{array}{ccccc} G & \xrightarrow{f} & G' & \xrightarrow{c} & \bar{G} \\ p\uparrow & & p'\uparrow & & \uparrow\bar{p} \\ A_0[X] & \xrightarrow{\phi} & B_0[X,Y] & \xrightarrow{\chi} & \bar{B}_0[Y] \end{array}$$

an embedding of $f : G \to G'$.

(i) For every flat local homomorphism $g_0 : (B_0, n_0) \to (B'_0, n'_0)$ of local rings satisfying

$$L(B'_0/n_0 B'_0) < \infty,$$

the composition $f' : G \to G' \to G' \otimes_{B_0} B'_0$ of f with the natural homomorphism $G' \to G' \otimes_{B_0} B'_0$, $g \mapsto g \otimes 1$ is a permissible graded map.

(ii) The diagram

(2)
$$
\begin{array}{ccccc}
G & \xrightarrow{f'} & G' \otimes_{B_0} B'_0 & \xrightarrow{c \otimes_{B_0} B'_0} & \bar{G} \otimes_{B_0} B'_0 \\
p \uparrow & & \uparrow p' \otimes_{B_0} B'_0 & & \uparrow \bar{p} \otimes_{B_0} B'_0 \\
A_0[X] & \xrightarrow{\phi'} & (B_0 \otimes_{B_0} B'_0)[X, Y] & \xrightarrow{\chi \otimes_{B_0} B'_0} & (\bar{B}_0 \otimes_{B_0} B'_0)[Y]
\end{array}
$$

obtained from (1) tensoring the square on the right with B'_0 over B_0 is an embedding of the permissible map f' of (i).

(iii) Assume that G is Artinian and that Y is finite. Then the Kodaira-Spencer maps of f and f' are related by the formula

$$Df' = Df \otimes_{B_0} B'_0.$$

The analogous formula is also valid for the embedded Kodaira-Spencer maps belonging to (1) and (2), respectively.

(iv) The Kodaira-Spencer map of the permissible graded map $f_{B_0} : G \otimes_{B_0} A_0 \to G'$ induced by $f : G \to G'$ coincides with the Kodaira-Spencer map of f,

$$Df_{B_0} = Df.$$

Proof. (i). Follows directly from the definition (cf 8.1). Note that $G' \otimes_{B_0} B'_0$ has finite Hilbert function by 2.8(i) since the special fibre of g_0 is assumed to be Artinian. Moreover, $\bar{G} \otimes_{B_0} B'_0$ is normally flat, since flatness is preserved under base change.

(ii). Follows from the definition of an embedding in 8.5 since a homogeneous generating system of G' over $G'(0)$ is also a homogeneous system of generators for $G' \otimes_{B_0} B'_0$ over $G'(0) \otimes_{B_0} B'_0$.

(iii). First consider the embedded case. Let $\bar{B}'_0 := B'_0/m_0 B'_0$. Since G is Artinian, hence a finitely generated module over A_0,

$$
\begin{aligned}
T_{G, \bar{B}'_0} &= \mathrm{Hom}_{A_0}(G^+/(G^+)^2, \bar{B}'_0) \\
&= \mathrm{Hom}_{B'_0}(G^+/(G^+)^2 \otimes_{A_0} B'_0, \bar{B}_0 \otimes_{B_0} B'_0) \\
&= \mathrm{Hom}_{B_0}(G^+/(G^+)^2 \otimes_{A_0} B_0, \bar{B}_0) \otimes_{B_0} B'_0 \\
&= T_{G, \bar{B}_0} \otimes_{B_0} B'_0
\end{aligned}
$$

Here the first identity is just the definition, the second comes from the universality property of the tensor product, and the third from the fact that the Hom-functor with a finitely presentable module in its first argument commutes with exact tensor products (see [Ma86], Th. 7.11). It is this latter identity where we need that G is Artinian.

Since Y is finite, the polynomial ring $\bar{B}_0[Y]$ is Noetherian. In particular, $\mathrm{Ker}(\bar{p})$ is a finitely generated ideal of $\bar{B}_0[Y]$. Therefore, letting $R := \bar{B}_0[Y]$ and $R' = (\bar{B}_0 \otimes_{B_0} B_0')[Y]$,

$$
\begin{aligned}
N_{\bar{p} \otimes_{B_0} B_0'} &= \mathrm{Hom}_{(\bar{B}_0 \otimes_{B_0} B_0')[X]}(\mathrm{Ker}(\bar{p}) \otimes_{B_0} B_0', \bar{G} \otimes_{B_0} B_0') \\
&= \mathrm{Hom}_{R'}(\mathrm{Ker}(\bar{p}) \otimes_R R', \bar{G} \otimes_R R') \\
&= \mathrm{Hom}_R(\mathrm{Ker}(\bar{p}), \bar{G}) \otimes_R R' \\
&= N_{\bar{p}} \otimes_{B_0} B_0'
\end{aligned}
$$

Here the first identity holds by definition, the second comes from the fact that $\mathrm{Ker}(\bar{p})$ and $\bar{G}$ are R-modules and that $R' = R \otimes_{B_0} B_0'$, the third uses that $\mathrm{Ker}(\bar{p})$ is finitely generated and R' is R-flat (see [Ma86], Th. 7.11), and the last results from the R-module property of $N_{\bar{p}} = \mathrm{Hom}_R(\mathrm{Ker}(\bar{p}), \bar{G})$. We have proved, the claim is true on the module level. We have to consider the mappings now. Compare the definitions (cf 8.5) of the embedded Kodaira-Spencer maps associated with the diagrams (1) and (2). One sees immediately that the latter one is the B_0'-linear extension of the former which is just the claim.

In order to prove statement (iii) in the non-embedded case, consider the exact sequences

$$
0 \to D_{\bar{p}} \to N_{\bar{p}} \to \mathrm{T}_{\bar{G}}^1 \to 0,
$$
$$
0 \to D_{\bar{p} \otimes_{B_0} B_0'} \to N_{\bar{p} \otimes_{B_0} B_0'} \to \mathrm{T}_{\bar{G} \otimes_{B_0} B_0'}^1 \to 0
$$

where

$$
D_{\bar{p}} := \mathrm{Der}_{\bar{B}_0}(\bar{B}_0[Y], \bar{G}),
$$
$$
D_{\bar{p} \otimes_{B_0} B_0'} := \mathrm{Der}_{\bar{B}_0 \otimes_{B_0} B_0'}((\bar{B}_0 \otimes_{B_0} B_0')[Y], \bar{G} \otimes_{B_0} B_0').
$$

To prove the claim, it will be sufficient to show that the tensor product with B_0' of right hand side homomorphism of the first sequence equals the right hand side homomorphism of the second. Since both sequences are exact, it will suffice to prove the analogous statement for the left hand side homomorphisms,

$$
(D_{\bar{p}} \to N_{\bar{p}}) \otimes_{B_0} B_0' = D_{\bar{p} \otimes_{B_0} B_0'} \to N_{\bar{p} \otimes_{B_0} B_0'}.
$$

Letting, as above, $R = \bar{B}_0[Y]$ and $R' = (\bar{B}_0 \otimes_{B_0} B_0')[Y]$,

$$
\begin{aligned}
D_{\bar{p} \otimes_{B_0} B_0'} &= \mathrm{Der}_{\bar{B}_0 \otimes_{B_0} B_0'}(R', \bar{G} \otimes_{B_0} B_0') \\
&= \mathrm{Hom}_{R'}(\Omega_{R'/\bar{B}_0 \otimes_{B_0} B_0'}, \bar{G} \otimes_{B_0} B_0') \\
&= \mathrm{Hom}_{R'}(\Omega_{R/\bar{B}_0} \otimes_R R', \bar{G} \otimes_R R') \\
&= \mathrm{Hom}_R(\Omega_{R/\bar{B}_0}, \bar{G}) \otimes_R R' \\
&= D_{\bar{p}} \otimes_{B_0} B_0'
\end{aligned}
$$

Here the first identity holds by definition, the second comes from the universality property of the module of Kähler differentials $\Omega_{R'/\bar{B}_0 \otimes_{B_0} B'_0}$ (see [Ma86], chapter 25), the third uses [Ma86], Ex. 25.4 and the fact that $\bar{G}$ is a module over R, and the fourth, [Ma86], Th. 7.11. Note that, by [Ma86] chapter 25, $\Omega_{R/\bar{B}_0}$ is a free module over $R = \bar{B}_0[X]$ of rank equal to the cardinality of Y. We have proved, the B'_0-module $D_{\bar{p} \otimes_{B_0} B'_0}$ is defined over B_0 with the defining submodule equal to $D_{\bar{p}}$. We have yet to show that the latter submodule is mapped into the defining submodule $N_{\bar{p}}$ of $N_{\bar{p} \otimes_{B_0} B'_0}$.

Consider the diagram

$$
\begin{array}{ccc}
D_{\bar{p}} & & N_{\bar{p}} \\
\| & & \| \\
\mathrm{Der}_{\bar{B}_0}(R, \bar{G}) & \longrightarrow & \mathrm{Hom}_R(\mathrm{Ker}(\bar{p}), \bar{G}) \\
\downarrow & & \downarrow \\
\mathrm{Der}_{\bar{B}_0 \otimes_{B_0} B'_0}(R \otimes_{B_0} B'_0, \bar{G} \otimes_{B_0} B'_0) & \longrightarrow & \mathrm{Hom}_{R \otimes_{B_0} B'_0}(\mathrm{Ker}(\bar{p}) \otimes_{B_0} B'_0, \bar{G} \otimes_{B_0} B'_0) \\
\| & & \| \\
D_{\bar{p} \otimes_{B_0} B'_0} & & N_{\bar{p} \otimes_{B_0} B'_0}
\end{array}
$$

where the horizontal maps are restriction to $\mathrm{Ker}(p)$ and $\mathrm{Ker}(\bar{p}) \otimes_{B_0} B'_0$, respectively, and the vertical homomorphisms are defined by B'_0-linear extension. Since B'_0-linear extension commutes with restriction, this diagram is commutative, as required. The proof of assertion (iii) is complete.

(iv). Follows from the fact that domain of definition

$$
T_{G, \bar{B}_0} := \mathrm{Hom}_{A_0}(G^+/(G^+)^2, \bar{B}_0)
$$

of the Kodaira-Spencer map can be identified with $\mathrm{Hom}_{B_0}(G^+_{B_0}/(G^+_{B_0})^2, \bar{B}_0)$ (see the beginning of 8.6).

8.10. Nice distinguished bases

Let (A_0, m_0) be an Artinian local ring, G a permissible graded algebra over A_0 (see 8.1), and $g := (g_\nu)_{\nu \in N}$ a family of homogeneous elements generating G as an algebra over A_0,

$$
G = A_0[g] = A_0[g_\nu | \nu \in N].
$$

We will refer to the family g as to a system of *homogeneous algebra generators* for G. Such a system is called *minimal*, if no proper subfamily of g generates G as an A_0-algebra. Note that for a subfamily g' of g to be minimal it is necessary and sufficient that for every d the degree d elements of g' form a minimal generating set of the A_0-module $G(d)/A_0[g_\nu | \deg g_\nu < d](d)$. Since G has finite Hilbert function (by 8.1), every system of homogeneous algebra generators for G contains a minimal subsystem.

Let $\mathbb{P}(g)$ denote the set of all power products of the elements g_ν. Then $\mathbb{P}(g)$ generates G as an A_0-module. A system of *monomial module generators* for G over g is by definition a family $g^* := (g^\nu)_{\nu \in N^*}$ of elements from $\mathbb{P}(g)$ generating G as a module over A_0. Note that, if $g = (g_\nu)_{\nu \in N}$ is minimal, then g^* contains g as a subfamily. In such a situation we will assume below, unless stated otherwise, that $N \subseteq N^*$ and $g_\nu = g^\nu$ for $\nu \in N$. A system g^* of monomial module generators over g is called *minimal*, if g is minimal and if no proper subfamily of g^* generates G as A_0-module. Note that, if g^* is a system of monomial module generators over g and g is minimal, then there is a subfamily of g^* which is a minimal system of monomial module generators over g. The argument for this is similar to the above one for the existence of minimal subsystems of g.

Let $f : (A, m) \rightarrow (B, n)$ be a homomorphism of filtered local rings with special fibre $\bar{B} := B/mB$ such that $\bar{G} := G(\bar{B})$ is normally flat. We will construct now a special type of distinguished basis for f which will be called a nice distinguished basis. By assumption, the algebra $\bar{G}$ is a permissible graded algebra over $\bar{G}(0) = \bar{B}(0)$ (see 8.1), and the homogeneous parts $\bar{G}(d)$ of $\bar{G}$ are finitely generated free modules over $\bar{G}(0)$ ([Ma86], Th. 7.10)). Choose a minimal system $\bar{g} := (\bar{g}_\nu)_{\nu \in N}$ of homogeneous algebra generators for $\bar{G}$ and a minimal system $\bar{g}^* := (\bar{g}^\nu)_{\nu \in N^*}$ of monomial module generators over $\bar{g}$. Further let $\overline{\varpi} := (\overline{\varpi}_\mu)_{\mu \in M}$ be a vector space basis of $\bar{G}(0)$ over $\bar{A} := A/m$. Then the family $(\overline{\varpi}_\mu \cdot \bar{g}^\nu)_{(\mu,\nu) \in M \times N}$ is a homogeneous vector space basis of $\bar{G}$ over $\bar{A}$. For every $\bar{g}_\nu \in \bar{G}$ select a representative in B, i.e., an element $b_\nu \in B$ such that $\mathrm{ord}_B b_\nu = \deg \bar{g}_\nu$ and

$$\bar{g}_\nu = \mathrm{in}(b_\nu \bmod mB) \quad (\in \bar{G}).$$

Let $b := (b_\nu)_{\nu \in N}$ denote the family of these lifts. For every power product $\bar{g}^\nu$ in the family $\bar{g}^*$ form the corresponding power product $b^\nu (\in B)$ of the elements in b. Then, for $\nu \in N$,

$$\mathrm{in}(b^\nu \bmod mB) = \bar{g}^\nu \in \bar{G} \text{ and } \mathrm{ord}_B b^\nu = \deg \bar{g}^\nu.$$

We are using the fact here that

$$\mathrm{in}(\beta)\,\mathrm{in}(\beta') = \mathrm{in}(\beta\beta') \quad \text{or} \quad \mathrm{in}(\beta)\,\mathrm{in}(\beta') = 0$$

for arbitrary $\beta, \beta' \in B$ and that the second case is impossible in our situation due to the choice of $\bar{g}^*$. For every $\mu \in M$ let $\omega_\mu \in B$ be a representative of $\overline{\varpi}_\mu \in \bar{G}(0)(= \bar{B}(0))$. We are now ready to define our distinguished basis. Let $\Lambda := M \times N$, and consider the mappings

$$\mathrm{b} : \Lambda \rightarrow B, \quad \mathrm{b}(\mu, \nu) := \omega_\mu \cdot b^\nu$$
$$\mathrm{d} : \Lambda \rightarrow \mathbb{N}, \quad \mathrm{d}(\mu, \nu) := \deg(\overline{\varpi}_\mu \cdot \bar{g}^\nu)(= \deg \bar{g}^\nu)$$

By construction,

$$\mathrm{ord}_B \mathrm{b}(\mu, \nu) = \mathrm{ord}_{\bar{B}}(\mathrm{b}(\mu, \nu) \bmod mB) = \deg(\overline{\varpi}_\mu \cdot \bar{g}^\nu) = \mathrm{d}(\mu, \nu)$$

and $\mathrm{in}(b(\mu,\nu) \bmod mB) = \overline{\omega}_\mu \cdot \overline{g}^\nu$. In particular, the elements $\mathrm{in}(b(\mu,\nu) \bmod mB)$ form a vector space basis of $\overline{G}$ over $\overline{A}$. If $\mathrm{d}(\mu,\nu) \geq k$ then $\mathrm{ord}_B\, b(\mu,\nu) \geq k$, i.e., $b(\mu,\nu) \in F_B^k$. Since F_B is an Artin-Rees filtration, $b(\mu,\nu) \in n^d$ for $k = k(d)$ sufficiently large (see 1.14(iii)). We have proved, the pair (b, d) is a distinguished basis for f. In that what follows, a distinguished basis obtained this way will be called a *nice distinguished basis* for f. Given a nice distinguished basis for f, we will refer to the associated families $\overline{g}$, $\overline{g}^*$, $\overline{\omega}$, b, b^*, ω used in the above construction as follows. The family $\overline{g}$ is called the *underlying system of algebra generators*, $\overline{g}^*$ the *underlying system of module generators*, $\overline{\omega}$ the *underlying vector space basis*. The families b, b^*, and ω will be called the *underlying lifts* of $\overline{g}$, $\overline{g}^*$, and $\overline{\omega}$, respectively. Here is a summary of the data related with the construction of a nice distinguished basis for $f : (A, m) \to (B, n)$.

underlying family	property
algebra generators $\overline{g}$ for $\overline{G} = G(B/mB)$	$\overline{G} = \overline{G}(0)[\overline{g}_\nu\,|\nu \in N^*]$
monomial module generators $\overline{g}^*$	$\overline{G} = \displaystyle\bigoplus_{\nu \in N} \overline{G}(0) \cdot g^\nu$
vector space basis $\overline{\omega}$	$\overline{G}(0) = \displaystyle\bigoplus_{\mu \in M} (A/m) \cdot \overline{\omega}_\mu$
lift b to B of $\overline{g}$	$\overline{g}_\nu = \mathrm{in}(b_\nu \bmod mB) \in \overline{G}$
monomial lift b^* to B of $\overline{g}^*$	$\overline{g}^\nu = \mathrm{in}(b^\nu \bmod mB) \in \overline{G}$
lift ω to B of $\overline{\omega}$	$\overline{\omega}_\mu = (\omega_\mu \bmod mB + F_B^1)$
generator function b	$b(\mu,\nu) := \omega_\mu \cdot b^\nu$
degree function d	$\mathrm{d}(\mu,\nu) := \deg(\overline{\omega}_\mu \cdot \overline{g}^\nu)$

Let $f_0 : (A_0, m_0) \to (B_0, n_0)$ be a flat local homomorphism of Artinian local rings and $f : G \to G'$ a permissible graded map over f_0. In this situation one can define various kinds of structure constants as follows. Let $\overline{g} := (\overline{g}_\nu)_{\nu \in N}$ be a minimal system of homogeneous algebra generators for $\overline{G}$, and $\overline{g}^* := (\overline{g}^\nu)_{\nu \in N^*}$ a minimal system of monomial module generators over $\overline{g}$. Since $\overline{G} := G'/(m, G^+)G'$ is normally flat by assumption, $\overline{g}^*$ is a free homogeneous generating system of $\overline{G}$ over $\overline{B}_0 := B_0/m_0 B_0 = \overline{G}(0)$ (see [Ma86], Th. 7.10). In particular, there are uniquely determined elements $\overline{b}^\gamma_{\alpha\beta} \in \overline{B}_0$ with

$$\overline{g}^\alpha \cdot \overline{g}^\beta = \sum_{\gamma \in N} \overline{b}^\gamma_{\alpha\beta} \cdot \overline{g}^\gamma.$$

We will refer to the elements $\overline{b}^\gamma_{\alpha\beta}$ as to the *structure constants* of $\overline{G}$ with respect to $\overline{g}^*$. Next, choose a family $g' := (g'_\nu)_{\nu \in N}$ of homogeneous lifts in G' of the elements from $\overline{g}$. For every power product in $\overline{g}^*$, form the corresponding power product of the lifts g'_ν. The family of power products $g'^* := (g'^\nu)_{\nu \in N^*}$ obtained this way will be called *monomial lift* to G' over g' of the monomial generating system $\bar{g}^*$. Note that by the graded Nakayama lemma 2.1 the elements g'^{ν} generate G' as a G_{B_0}-module, and that they are free over G_{B_0} by 2.7, since G' is G_{B_0}-flat (see also the second remark of 3.11). So there are uniquely determined homogeneous elements $g_{\alpha\beta}^{\nu} \in G_{B_0}$ satisfying

$$g'^{\alpha} \cdot g'^{\beta} = \sum_{\nu \in N} g_{\alpha\beta}^{\nu} \cdot g'^{\nu}.$$

We will refer to $g := (g_{\alpha\beta}^{\nu})_{\nu,\alpha,\beta \in N}$ as to the *family of structure constants* of G' over G_{B_0} relative to g'^*. Finally let $\overline{\omega} := (\overline{\omega}_{\mu})_{\mu \in M}$ be a vector space basis of $\bar{G}(0)$ over $\bar{A}_0 := A_0/m_0$. Then the uniquely determined elements $\bar{x}_{\alpha\beta}^{\mu} \in \bar{A}_0$ such that

$$\overline{\omega}_{\alpha} \cdot \overline{\omega}_{\beta} = \sum_{\mu \in M} \bar{x}_{\alpha\beta}^{\mu} \cdot \overline{\omega}_{\mu}$$

are called *structure constants* of $\bar{G}(0)$ over $\bar{A}_0$ with respect to $\overline{\omega}$. Lifting the generators $\overline{\omega}_{\mu}$ to $G'(0) = B_0$ we obtain a free generating system $\omega' := (\omega'_{\mu})_{\mu \in M}$ over $G(0) = A_0$, hence a uniquely determined family $x := (x_{\alpha\beta}^{\mu})_{\alpha,\beta,\mu \in M}$ of elements from A_0 with

$$\omega'_{\alpha} \cdot \omega'_{\beta} = \sum_{\mu \in M} x_{\alpha\beta}^{\mu} \cdot \omega'_{\mu}.$$

These are called *structure constants* of $G'(0)$ over $G(0)$ with respect to ω'. Here is a summary of the introduced data.

underlying family	property
algebra generators $\bar{g}$ for $\bar{G} = G'/(m, G^+)$	$\bar{G} = \bar{G}(0)[\bar{g}_{\nu} \mid \nu \in N^*]$
monomial module generators $\bar{g}^*$	$\bar{G} = \bigoplus_{\nu \in N} \bar{G}(0) \cdot \bar{g}^{\nu}$
structure constants $\bar{b}_{\alpha\beta}^{\gamma}$ of $\bar{G}$	$\bar{g}^{\alpha} \cdot \bar{g}^{\beta} = \sum_{\gamma \in N} \bar{b}_{\alpha\beta}^{\gamma} \cdot \bar{g}^{\gamma}$
lift g' to G' of $\bar{g}$	$\bar{g}_{\nu} = (g'_{\nu} \bmod(m, G^+))$
monomial lift g'^* to G' of $\bar{g}^*$	$\bar{g}^{\nu} = (g'^{\nu} \bmod(m, G^+))$ and
structure constants $g_{\alpha\beta}^{\lambda}$ of G'	$G' = \bigoplus_{\nu \in N} G_{B_0} \cdot g'^{\nu}$ $g'^{\alpha} \cdot g'^{\beta} = \sum_{\nu \in N} g_{\alpha\beta}^{\nu} \cdot g'^{\nu}$
basis $\overline{\omega}$ of $\bar{G}(0)$ over $\bar{A}_0 := A_0/m_0$	$\bar{B}_0 = \bigoplus_{\mu \in M} \bar{A}_0 \cdot \overline{\omega}_{\mu}$
structure constants $\bar{x}_{\alpha\beta}^{\mu}$ of $\bar{G}(0)$	$\overline{\omega}_{\alpha} \cdot \overline{\omega}_{\beta} = \sum_{\mu \in M} \bar{x}_{\alpha\beta}^{\mu} \cdot \overline{\omega}_{\mu}$
lift ω' to $G'(0)$ of $\overline{\omega}$	$B_0 = \bigoplus_{\mu \in M} A_0 \cdot \omega'_{\mu}$

structure constants $x_{\alpha\beta}^{\mu}$ of $G'(0)$ $\qquad$ $\Big|$ $\quad$ $\omega_{\alpha}' \cdot \omega_{\beta}' = \sum_{\mu \in M} x_{\alpha\beta}^{\mu} \cdot \omega_{\mu}'$

The next statement relates the structure constants of G' over G with those of G' over G_{B_0} in case both families of structure constants come from one and the same nice distinguished basis.

8.11. Structure constants associated with a nice distinguished basis

Let $f : (A, m) \to (B, n)$ be a homomorphism of complete filtered local rings and (b, d) a nice distinguished basis for f. Assume that the following conditions are satisfied.

1. f is tangentially flat.

2. $\bar{G} := G(B/mB)$ is normally flat.

Let

$$f' : G \to G' \quad \text{with } f' := G(f), G := G(A), G' := G(B)$$

be the homomorphism induced by f on the associated graded rings, and consider the following data (see 8.10 for details).

$\bar{g}^* = (\bar{g}^{\nu})_{\nu \in N^*}$ $\quad$ the underlying family of monomial module generators of (b, d)

$g'^* = (g'^{\nu})_{\nu \in N^*}$ $\quad$ a monomial lift to G' of $\bar{g}^*$

$(g_{\alpha\beta}^{\nu})$ $\quad$ the family of structure constants of G' over $G \otimes_{A(0)} B(0)$ with respect to g'^*

$\bar{\omega} = (\bar{\omega}_{\mu})_{\mu \in M}$ $\quad$ the underlying vector space basis of (b, d)

$\omega' = (\omega_{\mu}')_{\mu \in M}$ $\quad$ a lift to $G'(0) = B(0)$ of $\bar{\omega}$

$(x_{\alpha\beta}^{\mu})$ $\quad$ the family of structure constants of $G'(0)$ over $G(0) = A(0)$ with respect to ω'

$(a_{\alpha\beta}^{\mu})$ $\quad$ the family of structure constants of B over A with respect to (b, d) (see 5.7)

$(A_{\alpha\beta}^{\nu})$ $\quad$ the family of structure constants of G' over G defined by the family $(a_{\alpha\beta}^{\mu})$, i.e., $A_{\alpha\beta}^{\mu} := (a_{\alpha\beta}^{\mu} \bmod F^{d(\alpha)+d(\beta)-d(\mu)})$

Write

(1) $$g_{\alpha\beta}^{\nu} = \sum_{\mu \in M} g_{\alpha\beta\mu}^{0\nu} \otimes \omega_{\mu}' \quad \text{with} \quad g_{\alpha\beta\mu}^{0\nu} \in G$$

Then, for arbitrary $\alpha, \beta, \nu \in M$ and $\alpha', \beta', \nu' \in N^*$,

$$A_{(\alpha,\alpha')(\beta,\beta')}^{(\nu,\nu')} := \sum_{\mu,\mu' \in M} g_{\alpha'\beta'\mu}^{0\nu'} \cdot x_{\alpha\beta}^{\mu'} x_{\mu'\mu}^{\nu}$$

Proof. We will prove this by showing that the entries on the right are also structure constants with respect to $(\mathrm{in}(b(\lambda)))_{\lambda \in \Lambda := M \times N}$.

8.11

By definition, the structure constants of (b, d) satisfy

$$b(\alpha) \cdot b(\beta) = \sum_{\mu \in \Lambda} a^{\mu}_{\alpha\beta} \cdot b(\mu).$$

Since f is tangentially flat, $\operatorname{ord} a^{\mu}_{\alpha\beta} \geq \operatorname{ord} b(\alpha) + \operatorname{ord} b(\beta) - \operatorname{ord} b(\mu)$ (see 5.8). Therefore, taking initial forms on both sides, we obtain,

$$\operatorname{in}(b(\alpha)) \cdot \operatorname{in}(b(\beta)) = \sum_{\mu \in \Lambda} (a^{\mu}_{\alpha\beta} \bmod F^{\mathrm{d}(\alpha)+\mathrm{d}(\beta)-\mathrm{d}(\mu)}) \cdot \operatorname{in}(b(\mu,)).$$

In particular, the above defined elements $A^{\mu}_{\alpha\beta}$ are as stated the structure constants of G' over G with respect to $(b(\lambda))_{\lambda \in \Lambda}$.

On the other hand, the assumptions imply that f is a permissible graded homomorphism, hence by 8.10, the family g'^* is a free homogeneous generating set of the $G \otimes_{A(0)} B(0)$-module G', and the structure constants $g^{\nu'}_{\alpha'\beta'}$ are characterized by the identities

(2)
$$g'^{\alpha'} \cdot g'^{\beta'} = \sum_{\nu' \in N} g^{\nu'}_{\alpha'\beta'} \cdot g'^{\nu'}.$$

Since $G'(0) = B(0)$ is flat over $G(0) = A(0)$, the lifts ω'_{μ} form a free generating set of $B(0)$ over $A(0)$, hence the elements $1 \otimes \omega'_{\mu}$ form a free generating set of $G \otimes_{A(0)} B(0)$ over G. In particular, the structure constants $g^{\nu'}_{\alpha'\beta'}$ can be written in the indicated way (cf (1)). Substituting the right hand side of (1) into (2) and multiplying by $\omega'_{\alpha}\omega'_{\beta}$ we obtain

$$\omega'_{\alpha} g'^{\alpha'} \cdot \omega'_{\beta} g'^{\beta'} = \sum_{\nu' \in N, \mu \in M} g^{0\nu'}_{\alpha'\beta'\mu} \cdot \omega'_{\alpha}\omega'_{\beta}\omega'_{\mu} g'^{\nu'}.$$

Products of elements from ω' can be written as linear combinations of such elements using the structure constants $x^{\mu}_{\alpha\beta}$. Doing so, we get the identity,

$$\omega'_{\alpha} g'^{\alpha'} \cdot \omega'_{\beta} g'^{\beta'} = \sum_{\nu' \in N, \mu, \mu', \nu \in M} g^{0\nu'}_{\alpha'\beta'\mu} \cdot x^{\mu'}_{\alpha\beta} x^{\nu}_{\mu'\mu} \cdot \omega'_{\nu} g'^{\nu'}.$$

Now the elements $\omega'_{\mu} \cdot g'^{\mu'}$ form a free generating set of G' over G. In fact this is just the generating set defined by the distinguished basis (b, d),

$$\operatorname{in}(b(\mu, \mu')) = \operatorname{in}(\omega_{\mu} \cdot b^{\mu'}) = \omega_{\mu} \cdot g'^{\mu'}$$

(cf 5.3). In other words, the elements

$$X^{(\nu, \nu')}_{(\alpha, \alpha')(\beta, \beta')} := \sum_{\mu, \mu' \in M} g^{0\nu'}_{\alpha'\beta'\mu} \cdot x^{\mu'}_{\alpha\beta} x^{\nu}_{\mu'\mu}$$

120

are the structure constants of $G' = G(B)$ over $G = G(A)$ with respect to the family $(\mathrm{in}(\mathrm{b}(\mu,\mu')))_{\mu' \in N, \mu \in M}$, i.e.,

$$X^{(\nu,\nu')}_{(\alpha,\alpha')(\beta,\beta')} = A^{(\nu,\nu')}_{(\alpha,\alpha')(\beta,\beta')}.$$

as claimed.

8.12. The kernel of the Kodaira-Spencer map in low degree

Let $f_0 : (A_0, m_0) \to (B_0, n_0)$ be a flat local homomorphism of Artinian local rings with special fibre $\bar{B}_0$, $f : G \to G'$ a permissible graded map over f_0. Write

$$\bar{G} := G'/(m, G^+) \quad \text{and} \quad G_{B_0} := G \otimes_{A_0} B_0.$$

Further let

$\bar{g} := (\bar{g}_\nu)_{\nu \in N}$ be a minimal system of homogeneous algebra generators for $\bar{G}$

$\bar{g}^* := (\bar{g}^\nu)_{\nu \in N^*}$ a minimal system of monomial module generators for $\bar{G}$ over $\bar{g}$

$g'^* := (g'^\nu)_{\nu \in N^*}$ a monomial lift to G' of $\bar{g}^*$

$g := (g^\nu_{\alpha\beta})_{\nu,\alpha,\beta \in N}$ the structure constants of G' over G_{B_0} w.r.t. g'^*

Assume there is an integer k with

$$\deg \bar{g}_\nu \leq k \text{ for every } \nu \in N.$$

Then for a homogeneous element of degree $< -k$,

$$l \in \mathrm{Hom}_{B_0}(G^+_{B_0}/(G^+_{B_0})^2, \bar{B}_0),$$

the following conditions are equivalent.

(i) $l \in \mathrm{Ker}(Df)$, i.e., l is in the kernel of the Kodaira-Spencer map.

(ii) $l(g^\nu_{\alpha\beta} \mod (G^+_{B_0})^2) = 0$ for every structure constant $g^\nu_{\alpha\beta}$ of positive degree.

Proof. Step 1: Reduction to the embedded case. Let

$$
\begin{array}{ccccc}
G & \xrightarrow{f} & G' & \xrightarrow{c} & \bar{G} \\
{\scriptstyle p}\uparrow & & {\scriptstyle p'}\uparrow & & \uparrow{\scriptstyle \bar{p}} \\
A_0[X] & \xrightarrow{\phi} & B_0[X,Y'] & \xrightarrow{\chi'} & \bar{B}_0[Y']
\end{array}
\tag{1}
$$

be an embedding of f such that $Y' = (Y_\nu)_{\nu \in N'}$ has an indeterminate Y_ν for every module generator g'^ν of positive degree, i.e, $N' := N^* - \{0\}$ with 0 being the index of the unique generator g'^0 of degree 0, and such that p' is defined by

$$p'(Y_\nu) = g'^\nu \text{ for every } \nu \in N'.$$

8.12

Write as usual

$$I := \mathrm{Ker}(p), \quad I' := \mathrm{Ker}(p'), \quad \bar{I} := \mathrm{Ker}(\bar{p})$$

and

$$D' : T_{G,\bar{B}_0} := \mathrm{Hom}_{A_0}(G^+/(G^+)^2, \bar{B}_0) \to N_{\bar{g}^*} := \mathrm{Hom}_{B_0[Y']}(\mathrm{Ker}(\bar{p}), \bar{G})$$

for the embedded Kodaira-Spencer map associated with this embedding. Further let $g' := (g'_\nu)_{\nu \in N}$ be the family of homogeneous lifts in G' of $\bar{g}$ such that g'^* is a monomial lift over g', and let $Y := (Y_\nu)_{\nu \in N}$ be the subfamily of Y' consisting of the variables mapping to some element of the subfamily $g' \subseteq g'^*$. By assumption, each variable of Y has degree $\leq k$, and the subfamily Y defines also an embedding of f,

(2)
$$
\begin{array}{ccccc}
G & \xrightarrow{f} & G' & \xrightarrow{c} & \bar{G} \\
p\uparrow & & q'\uparrow & & \uparrow\bar{q} \\
A_0[X] & \xrightarrow{\phi} & B_0[X,Y] & \xrightarrow{\chi} & \bar{B}_0[Y]
\end{array}
$$

where q' and $\bar{q}$ are the restrictions of p' and $\bar{p}$ to the subrings $B_0[X,Y](\subseteq B_0[X,Y'])$ and $\bar{B}_0[Y](\subseteq \bar{B}_0[Y'])$, respectively. The Kodaira-Spencer map Df of f decomposes

$$Df : T_{G,\bar{B}_0} := \mathrm{Hom}_{A_0}(G^+/(G^+)^2, \bar{B}_0) \xrightarrow{D} N_{\bar{g}} := \mathrm{Hom}_{\bar{B}_0[Y]}(\mathrm{Ker}(\bar{q}), \bar{G}) \xrightarrow{r} T^1_{\bar{G}}$$

where D is the embedded Kodaira-Spencer map associated with the second embedding (2) and the kernel of the surjection r is equal to the image of the $\bar{B}_0[Y]$-linear map

$$s : \mathrm{Der}_{\bar{B}_0}(\bar{B}_0[Y], \bar{G}) \to N_{\bar{g}} = \mathrm{Hom}_{\bar{B}_0[Y]}(\mathrm{Ker}(\bar{q}), \bar{G})$$

induced by the inclusion $\mathrm{Ker}(\bar{q}) \subseteq \bar{B}_0[Y]$. The module of derivations on the left is generated by the partial derivatives $\frac{\partial}{\partial Y_\nu}$ with $\nu \in N$. Since $\deg Y_\nu \leq k$ for $\nu \in N$, these derivatives have degrees $\geq -k$. Therefore $\mathrm{Ker}(r) = \mathrm{Im}(s)$ is generated in degrees $\geq -k$ and r is an isomorphism in degrees $< -k$. But then condition (i) of the theorem is equivalent to $l \in \mathrm{Ker}(D)$, i.e., it will be sufficient to prove the claim with Df replaced by the embedded Kodaira-Spencer map D belonging to the embedding (2). In the second step we will reformulate the condition $l \in \mathrm{Ker}(D)$ in terms of the embedding (1).

Step 2: Proof of $l \in \mathrm{Ker}(Df) \Leftrightarrow l \in \mathrm{Ker}(D')$. By the choice of the families Y' and Y there exists for every $Y_\nu \in Y' - Y$ a power product $f_\nu(Y)$ with $Y_\nu - f_\nu(Y) \in \mathrm{Ker}(p')$. Consider the $B_0[X]$-algebra homomorphism $r : B_0[X,Y'] \to B_0[X,Y]$ that maps $Y_\nu \notin Y$ into $f_\nu(Y)$ and $Y_\nu \in Y$ into itself and define the $\bar{B}_0$-algebra homomorphism $\bar{r} : \bar{B}_0[Y'] \to \bar{B}_0[Y]$ in the same way. Then

$$
\begin{aligned}
r(\beta) &\equiv \beta \bmod \mathrm{Ker}(p') && \text{for every } \beta \in B_0[X,Y'], \\
\bar{r}(\beta) &\equiv \beta \bmod \mathrm{Ker}(\bar{p}) && \text{for every } \beta \in \bar{B}_0[Y'], \\
\chi \circ r &= \bar{r} \circ \chi'.
\end{aligned}
$$

The last identity means that the substitutions $Y_\nu \mapsto f_\nu(Y)$ commute with passage to residue classes. By the second congruence, an element of $\mathrm{Ker}(\bar{p})$ is mapped under $\bar{r}$ to an element of $\mathrm{Ker}(\bar{p}) \cap \bar{B}_0[Y] = \mathrm{Ker}(\bar{q})$, and the composition with $\bar{r}$ defines a homomorphism

$$\bar{r}^* : \mathrm{Hom}_{\bar{B}_0[Y]}(\mathrm{Ker}(\bar{q}), \bar{G}) \to \mathrm{Hom}_{\bar{B}_0[Y']}(\mathrm{Ker}(\bar{p}), \bar{G})$$

Moreover, $\bar{r}$ is left inverse to the natural inclusion $\bar{B}_0[Y] \subseteq \bar{B}_0[Y']$, hence $\bar{r}^*$ is right inverse to some map. In particular, $\bar{r}^*$ is injective. Consider the diagram

$$
\begin{array}{ccc}
\mathrm{Hom}_{A_0}(G^+/(G^+)^2, \bar{B}_0) & \xrightarrow{D'} & \mathrm{Hom}_{\bar{B}_0[Y']}(\mathrm{Ker}(\bar{p}), \bar{G}) \\
\| & & \uparrow \bar{r}^* \\
\mathrm{Hom}_{A_0}(G^+/(G^+)^2, \bar{B}_0) & \xrightarrow{D} & \mathrm{Hom}_{\bar{B}_0[Y]}(\mathrm{Ker}(\bar{q}), \bar{G})
\end{array}
$$

where D' and D are the embedded Kodaira-Spencer maps associated with the two given embeddings. In order to prove that this diagram is commutative, let $\bar{P} \in \mathrm{Ker}(\bar{p})$ and let $P'(X, Y') \in \mathrm{Ker}(p')$ be a lift of $\bar{P}$. Then $r(P') \in \mathrm{Ker}(q')$ is a lift of $\bar{r}(\bar{P}) \in \mathrm{Ker}(\bar{q})$ and

$$
\begin{aligned}
D_l(\bar{r}(\bar{P})) &= \bar{q}\left(\sum_{\gamma \in \Gamma} \frac{\partial r(P')(0, Y)}{\partial X_\gamma} \cdot l(p(X_\gamma) \bmod (G^+)^2)\right) \\
&= \bar{q}\bar{r}\left(\sum_{\gamma \in \Gamma} \frac{\partial P'(0, Y')}{\partial X_\gamma} \cdot l(p(X_\gamma) \bmod (G^+)^2)\right) \\
&= \bar{p}\left(\sum_{\gamma \in \Gamma} \frac{\partial P'(0, Y')}{\partial X_\gamma} \cdot l(p(X_\gamma) \bmod (G^+)^2)\right)
\end{aligned}
$$

hence $D_l(\bar{r}(\bar{P})) = D'_l(\bar{P})$ for every $\bar{P}$ hence $\bar{r}^*(D_l) = D'_l$. We have proved, the above diagram is commutative. Since r^* is injective, $l \in \mathrm{Ker}(Df) \Leftrightarrow l \in \mathrm{Ker}(D) \Leftrightarrow l \in \mathrm{Ker}(D')$.

Step 3: Construction of a generating system for I'. Let the elements $\bar{b}^\nu_{\alpha\beta} \in \bar{B}_0$ be the structure constant with respect to $\bar{g}^*$,

$$\bar{g}^\alpha \cdot \bar{g}^\beta = \sum_{\nu \in N'} \bar{b}^\nu_{\alpha\beta} \cdot \bar{g}^\lambda$$

Consider the polynomials

$$\bar{F}_{\alpha\beta}(Y') := Y_\alpha \cdot Y_\beta - \sum_{\nu \in N'} \bar{b}^\nu_{\alpha\beta} \cdot Y_\nu$$

$$F_{\alpha\beta}(X, Y') := Y_\alpha \cdot Y_\beta - \sum_{\nu \in N'} b^\nu_{\alpha\beta}(X) \cdot Y_\nu$$

where $b_{\alpha\beta}^\nu(X) \in B_0[X]$ is a homogeneous representative of $g_{\alpha\beta}^\nu \in G_{B_0}$. By construction, $\bar{F}_{\alpha\beta}(Y') \in \mathrm{Ker}(\bar{p})$ and $F_{\alpha\beta}(X, Y') \in \mathrm{Ker}(p')$. We want to prove that the polynomials $F_{\alpha\beta}(X, Y')$ together with the ideal I generate I',

$$(3) \qquad I' = (F_{\alpha\beta}(X, Y'), I | \alpha, \beta \in N').$$

Note that at least the right hand side of (3) is contained in the left. So there is a surjective homomorphism

$$C := B_0[X, Y']/(F_{\alpha\beta}(X, Y'), I | \alpha, \beta \in N') \to G',$$

and we have to prove that it is injective. Since the ring C is annihilated by I, it has the structure of a G-algebra and hence a G_{B_0}-algebra structure. Consider the G_{B_0}-submodule of C generated by the element $y_0 := g'^0 \in B_0$ and the residue classes y_ν of the indeterminates Y_ν,

$$V := \sum_{\nu \in N'} G_{B_0} \cdot y_\nu \subseteq C.$$

Since y_ν is mapped to g'^ν under the above surjection and the elements g'^ν form a free generating system of the G_{B_0}-module G', the restriction to V of our map is bijective, and it will be sufficient to show $V = C$. Now C is generated as a $B_0[X, Y']$-module by the degree zero element $g'^0 \in V$, so it is enough to show that V is a module over $B_0[X, Y']$. Since G_{B_0} has a $B_0[X]$-module structure, it suffices to show that V is invariant under multiplication by the indeterminates Y_α. But this is obvious: the products $Y_\alpha \cdot y_\beta$ can be expressed as G_{B_0}-linear combinations of the generators y_ν using the polynomials $F_{\alpha\beta}(X, Y')$ which have residue class zero in C. The proof of identity (3) is complete.

Step 4: the final proof. By the previous step $I' = (F_{\alpha\beta}(X, Y'), I | \alpha, \beta \in N')$, hence

$$\bar{I} = I' \cdot \bar{B}_0[Y'] = (\bar{F}_{\alpha\beta}(Y') | \alpha, \beta \in N').$$

Thus D_l' is zero if and only if it maps the polynomials $\bar{F}_{\alpha\beta}(Y')$ into zero. But

$$D_l'(\bar{F}_{\alpha\beta}(Y')) = \bar{p}\left(\sum_{\gamma \in \Gamma} \frac{\partial F_{\alpha\beta}(0, Y')}{\partial X_\gamma} \cdot l_*(X_\gamma) \right)$$

$$= -\bar{p}\left(\sum_{\gamma \in \Gamma} \sum_{\nu \in N'} \frac{\partial b_{\alpha\beta}^\nu(0)}{\partial X_\gamma} \cdot Y_\nu \cdot l_*(X_\gamma) \right)$$

$$= -\sum_{\nu \in N'} l_*\left(\sum_{\gamma \in \Gamma} \frac{\partial b_{\alpha\beta}^\nu(0)}{\partial X_\gamma} \cdot X_\gamma \right) \cdot \bar{g}^\nu$$

Here l_* is the composition $(X)B_0[X] \xrightarrow{p \otimes B_0} G_{B_0}^+ \to G_{B_0}^+/(G_{B_0}^+)^2 \xrightarrow{l} \bar{B}_0$, and l is identified with its B_0-linear extension $G_{B_0}^+/(G_{B_0}^+)^2 \to \bar{B}_0$ via the canonical isomorphisms

$$\text{Hom}_{A_0}(G^+/(G^+)^2, \bar{B}_0) \cong \text{Hom}_{B_0}(G^+/(G^+)^2 \otimes_{A_0} B_0, \bar{B}_0)$$
$$\cong \text{Hom}_{B_0}(G^+_{B_0}/(G^+_{B_0})^2, \bar{B}_0).$$

Note that $\bar{p}(Y_\nu) = \bar{g}^\nu$. Since the generators $\bar{g}^\nu$ are linearly independent over $\bar{B}_0$, l is in the kernel of D' if and only if all the coefficients in the last sum above are zero. If the element $b^\nu_{\alpha\beta}(X) \in B_0[X]$ is of degree zero, all its derivatives are zero and the condition is automatically satisfied (since each X_γ has positive degree by 8.5). So we may restrict to the elements $b^\nu_{\alpha\beta}(X)$ which have positive degrees. For these the condition is

$$0 = l_* \left(\sum_{\gamma \in \Gamma} \frac{\partial b^\nu_{\alpha\beta}(0)}{\partial X_\gamma} \cdot X_\gamma \right) = l_* \left(b^\nu_{\alpha\beta}(X) \right) = l(g^\nu_{\alpha\beta} \bmod (G^+)^2)$$

which is just condition (ii) of the theorem. Note that we use the fact that $b^\nu_{\alpha\beta}(X)$, as a homogeneous polynomial, has free member zero and that all non-linear terms are in the kernel of l_*.

8.13. Injectivity of the Kodaira-Spencer map

Let $f : (A, m) \to (B, n)$ be a flat homomorphism of complete filtered local rings with special fibre $\bar{B} := B/mB$ and (b, d) a nice distinguished basis for f. Assume that the following conditions are satisfied.

1. $G(B)$ is normally flat.
2. F_B is F_A-minimal with respect to (b, d).
3. The underlying system $\bar{g} := (\bar{g}_\nu)_{\nu \in N}$ of algebra generators for (b, d) consists of elements with restricted degrees, say $\deg \bar{g}_\nu \leq k$ for every $\nu \in N$.

Define

$$F := F(f) \quad \text{(the minimal flatifying prefiltration for } f)$$
$$I := G^+_{F_A}(A)$$
$$G := G_F(A)/(I) \quad (= G_F(A)/I \cdot G_F(A))$$
$$G' := G_{F_B + f_* F}(B)/(I)$$

Then the induced map

$$\phi := G_F(f) \otimes_{G_F(A)} G : G \to G'$$

is a permissible graded homomorphism, and the restriction to

$$T_{G,K} := \text{Hom}_A(G^+/(G^+)^2, K), \quad K := A/m,$$

of the Kodaira-Spencer map of ϕ,

$$D\phi : T_{G,\bar{B}(0)} := \mathrm{Hom}_A(G^+/(G^+)^2, \bar{B}(0)) \to T := \mathrm{T}^1_{G(\bar{B})}$$

is injective in degrees $< -k$. In particular, the Kodaira-Spencer map $D\phi$ itself is injective in degrees $< -k$, if it is definable over K with the defining subspace of $T_{G,\bar{B}(0)}$ equal to $T_{G,K}$ (which is trivially the case when $K = \bar{B}(0)$).

Proof. By definition, F is flatifying for f, i.e., $\mathrm{G}_{F_B + f_* F}(B)$ is flat over $\mathrm{G}_F(A)$. In particular,

$$G'/(m, G^+)G' = \mathrm{G}_{F_B + f_* F}(B)/(\mathrm{in}(m)) \cong \mathrm{G}_{F_B + f_* F}(B/mB) = \mathrm{G}_{F_B}(B/mB)$$

is normally flat by assumption 1, hence ϕ is a permissible graded homomorphism (cf 8.1). Let $\bar{g} := (\bar{g}_\nu)_{\nu \in N}$, $\bar{g}^* := (\bar{g}^\nu)_{\nu \in N^*}$, and $\bar{\omega} := (\bar{\omega}_\mu)_{\mu \in M}$ be as in 8.10 the underlying system of algebra generators, module generators, and the underlying vector space basis for the given nice distinguished basis (b, d) and let g', g'^*, and $\omega := (\omega_\mu)_{\mu \in M}$ be the respective underlying lifts to G' of $\bar{g}$, $\bar{g}^*$, and $\bar{\omega}$. Further let $g := (g^\nu_{\alpha\beta})_{\nu, \alpha, \beta \in N^*}$ be the family of structure constants of G' over $G_B := G \otimes_{A(0)} B(0)$ relative to g'^*. Then by 8.12, a homogeneous element l of degree $< -k$ is in the kernel of $D\phi$ if and only if

$$(1) \qquad\qquad l(g^\nu_{\alpha\beta} \bmod(G_B^+)^2) = 0$$

for arbitrary α, β, ν such that $g^\nu_{\alpha\beta}$ has positive degree. Using the terminology of 8.11, condition (1) reads

$$0 = l\big(\sum_{\mu \in M} g^{0\nu}_{\alpha\beta\mu} \bmod(G^+)^2\big) \cdot \bar{\omega}_\mu.$$

If l is in the subspace $T_{G,K} \subseteq T_{G,\bar{B}(0)}$, i.e., if it takes values in K, the latter implies, since the elements $\bar{\omega}_\mu$ are linearly independent over K,

$$0 = l\big(\sum_{\mu \in M} g^{0\nu}_{\alpha\beta\mu} \bmod(G_{B_0}^+)^2\big).$$

By 8.11, $l(A^\lambda_{\alpha\beta} \bmod(G^+)^2) = 0$.

Now by 7.6, the algebra $\mathrm{G}_F(A)$ is generated over $\mathrm{G}_{F_A}(A)$ by the elements $A^\lambda_{\alpha\beta}$, hence the irrelevant ideal G^+ is generated by those of the elements $A^\lambda_{\alpha\beta}$ that have positive degrees. Note that we have factored out the irrelevant ideal I of $\mathrm{G}_{F_A}(A)$. As a module over $\mathrm{G}_{F_A}(A)/I = \mathrm{G}_{F_A}(A)(0)$, $G^+/(G^+)^2$ is generated by the residue classes of these $A^\lambda_{\alpha\beta}$. But then the above identities imply that $l \in T_{G,K}$ is zero. This proves the claim.

9. Inequalities related with flat couples of local rings

9.1. A Lech type inequality

Let $f : (A, m) \to (B, n)$ be a flat homomorphism of filtered local rings and F a flatifying filtration for f with $F^1 = F_A^1$. Then

$$H_A^1 \cdot H_{B/mB}^0 \leq H_B^1 \cdot \prod_{d=2}^{\infty} \left(\frac{1 - T^d}{1 - T} \right)^{n(d)}$$

where $n(d) := \mu_A(M_F(d))$, $M_F := G_F^+(A)/(G_F^+(A)^2 + G_{F_A}^+(A) \cdot G_F(A))$.

Proof. It is sufficient to show that for every $N \in \mathbb{N}$ the inequality holds in degrees $\leq N$. Replacing f by $f \otimes_A A/F_A^{N+1}$, we may assume that A has finite length. In particular,

$$L := \mu_A(M_F(> 1))$$

is finite. The further proof precedes by induction on L. Let $L = 0$. It will be sufficient to show that $f : (A, m) \to (B, n)$ is tangentially flat, for, this implies $H_B^1 = H_A^1 \cdot H_{B/mB}^0$ (see 6.13) so that the inequality to be proved is trivial. To prove that f is tangentially flat, we have only to show that F_A is equal to the flatifying filtration F (see 7.1),

$$F_A^d = F^d \quad \text{for every } d.$$

For $d = 1$ this is true by assumption, so let $d > 1$. Since $L = 0$,

$$0 = M_F(> 1) = G_F^+(A)(> 1)/(G_F^+(A)^2 + G_{F_A}^+(A)(> 1) \cdot G_F(A)(0)).$$

Using this identity and the induction hypothesis with respect to d, we see that

$$
\begin{aligned}
F^d/F^{d+1} = G_F(A)(d) &= G_F^+(A)^2(d) + G_{F_A}^+(A)(d) \cdot G_F(A)(0) \\
&= \sum_{i,j>0, i+j=d} G_F^+(A)(i) \cdot G_F^+(A)(j) + (F_A^d + F^{d+1})/F^{d+1} \\
&= \sum_{i,j>0, i+j=d} F^i/F^{i+1} \cdot F^j/F^{j+1} + (F_A^d + F^{d+1})/F^{d+1} \\
&= (F_A^d + F^{d+1})/F^{d+1}.
\end{aligned}
$$

Therefore, $F^d = F_A^d + F^{d+1}$ hence $F^d = F_A^d + F^k$ for every $k \geq d$. Since F is an Artin-Rees filtration,

$$F^k \subseteq (F^1)^d = (F_A^1)^d \subseteq F_A^d$$

for large k (see 1.14(iii)), i.e., $F^d = F_A^d$ for every d. We have proved the claim for $L = 0$. Now let $L > 0$. Choose any element $x \in A$ that represents a homogeneous generator of order $l_0 > 1$ of the A-module $M_F(> 1)$. We may assume that $\operatorname{ord}_F(x) = l_0$. Consider the commutative diagram of filtered local rings,

$$
\begin{array}{ccc}
(A, m) & \xrightarrow{f} & (B, n) \\
\alpha \downarrow & & \downarrow \beta \\
(A', m') & \xrightarrow{f'} & (B', n')
\end{array} ,
$$

with

$$
A' := A[[X \mid \operatorname{ord} X = 1]]/(X^{l_0} - x)
$$
$$
B' := B[[X \mid \operatorname{ord} X = 1]]/(X^{l_0} - x)
$$

(see 6.6). Further let F' be the filtration $F' := F \cdot A' + (X)$ where (X) stands for the (X)-adic filtration of A', i.e.,

$$
F'^d := \sum_{i+j=d} F^i X^j A'.
$$

We want to apply the induction hypothesis (with respect to L) to the homomorphism $f' : (A', m') \to (B', n')$ and the filtration F'. So we have to check whether these data satisfy the conditions of the theorem to be proved. Since $\alpha : A \to A'$ and $\beta : B \to B'$ are finite extensions, the above diagram is Cartesian, $A' \otimes_A B \cong B'$, hence f' is flat. Since F is a flatifying filtration for f,

$$
\mathrm{H}^1_{(B, F_B + f_* F)} = \mathrm{H}^1_{(A, F)} \cdot \mathrm{H}^0_f .
$$

By 6.9,

$$
\mathrm{H}^1_{(A', F')} \geq \mathrm{H}^1_{(A; F)} \cdot \frac{1 - T^{l_0}}{1 - T} = \mathrm{H}^1_{(A; F)} \cdot \mathrm{H}^0_\alpha
$$

and, similarly,

$$
\mathrm{H}^1_{(B', F_{B'} + f'_* F')} \geq \mathrm{H}^1_{(B, F_B + f_* F)} \cdot \frac{1 - T^{l_0}}{1 - T} = \mathrm{H}^1_{(B, F_B + f_* F)} \cdot \mathrm{H}^0_\beta
$$

Note that $F_{B'} + f'_* F' = F_B B' + (X) + (F A') B' = (F_B + f_* F) B' + (X)$. Since the converse inequalities are also satisfied (cf 6.10), we have equality in the above estimations. But then

$$
\mathrm{H}^1_{(B', F_{B'} + f'_* F')} = \mathrm{H}^1_{(B, F_B + f_* F)} \cdot \frac{1 - T^{l_0}}{1 - T} = \mathrm{H}^1_{(A, F)} \cdot \mathrm{H}^0_f \cdot \frac{1 - T^{l_0}}{1 - T} = \mathrm{H}^1_{(A', F')} \cdot \mathrm{H}^0_{f'} .
$$

The homomorphism $(A', F') \to (B', F_{B'} + f_* F')$ is tangentially flat, i.e., F' is a flatifying filtration for $f' : (A', m') \to (B', n')$. Note that $F' = F \cdot A' + (X)$

is an Artin-Rees filtration by 1.15. Thus f' and F' satisfy the conditions of the theorem. Define

$$L' := \mu_{G_{F'}(A')}(M_{F'}(> 1)) \text{ with}$$
$$M_{F'} := G_{F'}^+(A')/(G_{F'}^+(A')^2 + G_{F_{A'}}^+(A') \cdot G_{F'}(A')).$$

We want to show that $L' < L$. Since x has F-order l_0, $X^{l_0} - \text{in}(x)$ is a homogeneous polynomial in $G_F(A)[X]$ that is not a zero-divisor. Therefore, up to isomorphism,

$$G_{F'}(A') = G_F(A)[X]/(X^{l_0} - \text{in}(x)) = \sum_{i=0}^{l_0-1} G_F(A) \cdot X^i$$

Moreover, since $x \in F^1 = F_A^1$, there are natural isomorphisms

$$G_{F_{A'}}(A')/(G_{F_A}^+(A), X) \cong G_{F_A}(A)[X]/(G_{F_A}^+(A), X, \text{in}(X^{l_0} - x))$$
$$\cong G_{F_A}(A)(0) \cong G_{F_{A'}}(A')(0)$$
$$\cong G_{F_{A'}}(A')/G_{F_{A'}}^+(A')$$

hence

$$G_{F_{A'}}^+(A') \cdot G_{F'}(A') = (G_{F_A}^+(A), X)$$

Identifying, as above, $G_{F'}(A')$ with a direct sum of copies of $G_F(A)$, one sees,

$$G_{F'}^+(A') = G_F^+(A) + \sum_{i=1}^{l_0-1} G_F(A) \cdot X^i$$

$$G_{F'}^+(A')^2 = G_F^+(A)^2 + \text{in}(x) G_F(A) + G_F^+(A) \cdot X + \sum_{i=2}^{l_0-1} G_F(A) \cdot X^i$$

hence

$$M_{F'} = G_{F'}^+(A')/(G_{F'}^+(A')^2 + G_{F_{A'}}^+(A') \cdot G_{F'}(A'))$$
$$= G_F^+(A)/(G_F^+(A)^2 + G_{F_A}^+(A) \cdot G_F(A) + \text{in}(x) G_F(A))$$
$$= M_F/x' G_F(A)$$

where x' denotes the residue class in M_F of $\text{in}(x)$. Since $\text{in}(x)$ represents a generator of $M_F(> 1)$, $L' < L$. The induction hypothesis applies to f' and F' giving the estimation

$$H_{A'}^1 \cdot H_{f'}^0 \le H_{B'}^1 \cdot \prod_{d=2}^{\infty} \left(\frac{1 - T^d}{1 - T}\right)^{n'(d)}$$

with

$$n'(d) = \mu_A(M_{F'}(d)) = \begin{cases} n(d) & \text{if } d \neq l_0 \\ n(d) - 1 & \text{if } d = l_0 \end{cases}$$

Since $H^0_{f'} = H^0_{B'/m'B'} = H^0_{B/mB} = H^0_f$, the proof of the theorem will be complete if we can show that

$$H^1_A \leq H^1_{A'} \quad \text{and} \quad H^1_{B'} \leq H^1_B \cdot \frac{1 - T^{l_0}}{1 - T}.$$

By construction, $x \in F^1 = F^1_A$, hence the polynomial $X^{l_0} - x$ has order at least 1 in the power series ring $A[[X \mid \operatorname{ord} X = 1]]$, i.e., the first inequality follows from 6.9. As for the second inequality, note that by 6.10

$$H^1_{B'} \leq H^1_B \cdot H^0_\beta = H^1_B \cdot \frac{1 - T^{l_0}}{1 - T}.$$

9.2. Main Theorem

Let $f : (A, m, K) \to (B, n, L)$ be a residually separable flat local homomorphism of local rings. Equip the rings A and B with their respective natural filtrations. Then

$$H^1_A \cdot H^0_{B/mB} \leq H^1_B \cdot \prod_{d=2}^{\infty} \left(\frac{1 - T^d}{1 - T} \right)^{n(d)}$$

with $n(d) := \dim_L T^1_{G(B/mB)}(-d)$.

Proof. Step 1: Reduction to the case that the field extension $K \subseteq L$ induced by f is trivial, i.e., that f is residually rational. We may assume that A and B are complete local rings. Our aim in this step is to find a decomposition $(A, m) \to (A', m') \to (B, n)$ of f into flat local homomorphisms such that the first factor satisfies $mA' = m'$ and second induces an isomorphism $A'/m' \to B/n$. This will reduce the proof of our inequality to the corresponding inequality for the second factor $(A', m') \to (B, n)$ of the decomposition.

By Cohen structure theory ([G-D71], EGA IV_1, §19) there are Cohen rings C_A, C_B (see [G-D71], (19.8.4)) and local homomorphisms $C_A \to A$ and $C_B \to B$ inducing isomorphisms of the residue class fields (see [G-D71], (19.8.6)(ii) and (i)). The two maps fit into a commutative diagram

$$
\begin{array}{ccc}
A & \longrightarrow & B/n \\
\uparrow & & \uparrow \\
C_A & & C_B \\
\uparrow & \nearrow & \\
\mathbb{Z}_{(\pi)} & &
\end{array}
$$

where the upper horizontal map is the composition of f with the canonical homomorphism $B \to B/n$ and $\mathbb{Z}_{(\pi)}$ is the localization of the integers $\mathbb{Z}$ at the characteristics π of A/m in case this characteristics is positive. In the characteristics zero case let denote $\mathbb{Z}_{(\pi)}$ the rational numbers. The maps from $\mathbb{Z}_{(\pi)}$ to C_A and C_B are the uniquely determined local homomorphisms preserving the identity element. Since C_A, as a Cohen ring, is formally smooth over $\mathbb{Z}_{(\pi)}$ (see [G-D71], (19.7.1) and (19.6.1)), there is a homomorphism $C_A \to C_B$ that can be commutatively included into this diagram, hence there is a commutative diagram of local homomorphisms

$$
\begin{array}{ccc}
C_B & \longrightarrow & B/n \\
\uparrow & & \uparrow \\
C_A & \to \quad A \xrightarrow{\ j\ } & B
\end{array}
$$

Since C_B is residually separable over C_A (for f has this property by assumption), the homomorphism $C_A \to C_B$ is formally smooth (by [G-D71], (19.7.1) and (19.6.1)), hence there is a local homomorphism $C_B \to B$ that can be commutatively included into the diagram. Adjoining a set of variables X to C_A that map to generators of the maximal ideal of A, the homomorphism $C_A \to A$ can be extended to a surjection $p_A : C_A[[X]] \to A$. Similarly one obtains a surjection $p_B : C_B[[X,Y]] \to B$ fitting into a commutative diagram

$$
\begin{array}{ccc}
A & \xrightarrow{\ j\ } & B \\
p_A \uparrow & & \uparrow p_B \\
C_A[[X]] & \xrightarrow{\ \varphi\ } & C_B[[X,Y]]
\end{array}
$$

where X and Y are finite sets of indeterminates, and φ induces the identity map on X. Compared with the diagram constructed in the proof of Cohen factorization theorem 5.9 the present diagram has the additional property (due to the separability assumption on f) that $\varphi(C_A) \subseteq C_B$. In particular, φ factors over $C_B[[X]]$. Define

$$
I := \mathrm{Ker}(p_A)
$$

and use p_A to identify A with $C_A[[X]]/I$. Then, tensoring the bottom row of the diagram with A over $C_A[[X]]$ yields a decomposition of f into local homomorphisms

$$
f : A = C_A[[X]]/I \to C_B[[X]]/(I) \to C_B[[X,Y]]/(I) \to B.
$$

where the first one is induced by the homomorphism $C_A \to C_B$ (hence is flat) and the last is induced by p_B. Define

$$
A' := C_B[[X]]/(I).
$$

Then A' is a local ring, flat over A with maximal ideal $m' := mA'$. By construction, the homomorphism $A' \to B$ is residually rational. In order to prove that B is flat over A' consider the canonical homomorphism

$$
G(A') \otimes_{A'} B \to G(B)
$$

induced by multiplication where A' and B are equipped with the m'-adic filtration. By the local flatness criterion ([Ma86], Th. 22.3(4')), it is sufficient to show that this homomorphism is bijective. Since $m' = mA'$, the m'-adic filtration of B coincides with the m-adic one. The fact that A' is A-flat, gives raise of a canonical isomorphism

$$G(A) \otimes_A A' \to G(A'),$$

hence it is sufficient to show that the canonical homomorphism

$$G(A) \otimes_A B = G(A) \otimes_A A' \otimes_{A'} B \to G(B),$$

is bijective, which happens to be the case since B is A-flat.

Step 2: Proof of the claim in case f is residually rational. We assume, as above, that A and B are complete local rings. Replacing f by $f \otimes_A A/m^{d+1}$ for increasing d, we may additionally assume that A is an Artinian local ring. Let F be the minimal flatifying filtration for f. Note that F is an Artin-Rees filtration, since the conditions of 7.9 are satisfied when the filtrations are the natural ones. Since F_A^1 is the maximal ideal of A and $F_A \subseteq F$, $F^1 = F_A^1$. By the previous theorem 9.1,

$$\mathrm{H}_A^1 \cdot \mathrm{H}_{B/mB}^0 \leq \mathrm{H}_B^1 \cdot \prod_{d=2}^{\infty} \left(\frac{1 - T^d}{1 - T} \right)^{n'(d)}$$

where $n'(d) := \mu_A(M_F(d))$ with $M_F := \mathrm{G}_F^+(A)/(\mathrm{G}_F^+(A)^2 + \mathrm{G}_{F_A}^+(A) \cdot \mathrm{G}_F(A))$. So it will be sufficient to show that for every $d > 1$,

$$n'(d) \leq n(d) := \dim_L \mathrm{T}_{\mathrm{G}(B/mB)}^1(-d).$$

Note that

$$
\begin{aligned}
n'(d) &:= \mu_A(M_F(d)) \\
&= \dim_K M_F(d)/mM_F(d) \\
&= \dim_L \mathrm{Hom}_A(M_F/mM_F, L)(-d) \\
&= \dim_L \mathrm{Hom}_A(M_F, L)(-d).
\end{aligned}
$$

To prove the claim, it will be sufficient to show that there is a graded L-linear map

$$\mathrm{Hom}_A(M_F, L) \to \mathrm{T}_{\mathrm{G}(B/mB)}^1$$

which is injective in all degrees < -1. Since f is residually rational, such a map is the Kodaira-Spencer map of

$$(\mathrm{G}_F(A) \to \mathrm{G}_{F_B + f_* F}(B)) \otimes_{\mathrm{G}_{F_A}(A)} \mathrm{G}_{F_A}(A)/\mathrm{G}_{F_A}^+(A)$$

(by 8.13). Note that the conditions of 8.13 are satisfied with $k = 1$. For, by assumption, $F_A \cdot B/mB$ is the natural filtration, i.e., the algebra $\mathrm{G}(B/m)$ is generated in degree 1 and the third condition of 8.13 holds with $k = 1$ for any nice distinguished basis (b, d) for f. As for the second condition, let I denote the ideal generated by all non-units among the elements $b(\lambda)$. Then $n = I + mB$, since I generates the maximal ideal in B/mB, hence for every d,

$$F_B^d = n^d = m^d I^0 + m^{d-1} I^1 + m^{d-2} I^2 + \ldots + m^0 I^d$$

$$= \sum_{i=0}^{d} F_A^{d-i} I^i,$$

i.e., F_B is F_A-minimal with respect to (b, d) (cf the definition in 5.4). The normal flatness condition on $G(B/mB)$ is trivially satisfied since the degree zero part of $G(B/mB)$ is the field L.

9.3. Germs in 3-space defined by monomials of degrees ≤ 3

The table below lists all ideals I of the polynomial ring $K[x, y, z]$ generated by power products of degrees at most three such that $B_0 := K[x, y, z]/I$ is an Artinian local ring (up to permutation of the indeterminates, K being a finite field of the Macaulay characteristics 31991). We calculate, using the Bayer-Stillman Computer algebra system Macaulay, the Hilbert series $\mathrm{H}_{B_0}^0$, the Hilbert series of the normal module $\mathrm{Hom}_{B_0}(I/I^2, B_0)$, and the Product, associated with Schlessinger's T^1,

$$P := \prod_{d=2}^{\infty} ((1 - T^d)/(1 - T))^{n(d)},$$

called *exponential Hilbert series* below. Note that, if the Hilbert series $\mathrm{H}_{B_0}^0$ is greater than ore equal to the exponential one, the Main Theorem 9.2 gives us an estimation $\mathrm{H}_A^1 \cdot P \leq \mathrm{H}_A^1 \cdot \mathrm{H}_{B/mB}^0 \leq \mathrm{H}_B^1 \cdot P$. Multiplying with power series of type $(1 - T^d)^{-1}$, which have non-negative coefficients, one obtains an inequality between certain sum transforms of H_A^1 and H_B^1, i.e., Lech's problem has an affirmative answer. This is the case for the first 78 singularities below, except for example 55, which is, however, a complete intersection. Since Lech's inequality always holds for complete intersections (see, e.g., [He90]), one can add Examples 55, 102, 109, 111, 114, and 115 to the list of singularities admitting a positive answer to Lech's problem. We formulate this result as a corollary.

Corollary

Let $f : (A, m, K) \rightarrow (B, n, L)$ be a residually separable flat local homomorphism of local rings such that B/mB is isomorphic to the factor ring associated to one of the examples 1 to 78, or 102, 109, 111, 114, 115 below. Then there is a positive integer i such that the Hilbert series of A and B (with respect to the natural filtrations) satisfy

$$\mathrm{H}_A^i \leq \mathrm{H}_B^i$$

9.4. Table

The ideals of $K[x,y,z]$ generated by power products of degrees two and three (K the finite field with 31991 elements, the default characteristic of Macaulay):

No.	Generators	Hilbert series	Hilbert series of the normal module	exponential Hilbert series
1.	$x^3, x^2y, xz^2, y^3, y^2z,z^3$	$1 +3T +6T^2 +4T^3$	$24T^{-1}+24$	1
2.	$x^3,x^2y,xy^2,xyz, xz^2,y^3,z^3$	$1+3T+6T^2+3T^3 +T^4$	$21T^{-1}+14+7T$	1
3.	$x^3,x^2y,x^2z,xy^2, xz^2,y^3,z^3$	$1+3T+6T^2+3T^3 +T^4$	$21T^{-1}+14+7T$	1
4.	$x^3,x^2y,xyz,xz^2, y^3,y^2z,z^3$	$1 +3T +6T^2 +3T^3$	$24T^{-1}+21$	1
5.	$x^3,x^2y,x^2z,xy^2, xyz,xz^2,y^3,z^3$	$1+3T+6T^2+2T^3 +T^4$	$31T^{-1}+6+8T$	1
6.	$x^3,x^2y,x^2z,xy^2, xz^2,y^3,y^2z,z^3$	$1 +3T +6T^2 +2T^3$	$30T^{-1}+16$	1
7.	$x^3,x^2y,x^2z,xyz, y^3,y^2z,yz^2,z^3$	$1 +3T +6T^2 +2T^3$	$30T^{-1}+16$	1
8.	$x^3,x^2y,x^2z,xy^2, xyz,y^3,yz^2,z^3$	$1 +3T +6T^2 +2T^3$	$30T^{-1}+16$	1
9.	$x^3,x^2y,x^2z,xy^2, xz^2,y^3,y^2z,yz^2, z^3$	$1+3T+6T^2+T^3$	$42T^{-1}+9$	1
10.	$x^3,x^2y,x^2z,xy^2, xyz,xz^2,y^3,y^2z, z^3$	$1+3T+6T^2+T^3$	$42T^{-1}+9$	1
11.	$x^3,x^2y,x^2z,xy^2, xyz,xz^2,y^3,y^2z, yz^2,z^3$	$1+3T+6T^2$	$60T^{-1}$	1
12.	$x^2, xy^2, xyz, y^3, yz^2,z^3$	$1 +3T +5T^2 +2T^3$	$16T^{-1}+15+2T$	1
13.	$x^2, xy^2, y^3, y^2z, yz^2,z^3$	$1 +3T +5T^2 +2T^3$	$16T^{-1}+15+2T$	1
14.	$x^2, xyz, y^3, y^2z, yz^2,z^3$	$1 +3T +5T^2 +2T^3$	$16T^{-1}+15+2T$	1
15.	$x^2, xy^2, xyz, y^3, y^2z,yz^2,z^3$	$1+3T+5T^2+T^3$	$24T^{-1}+11+T$	1
16.	$x^2, xy^2, xz^2, y^3, y^2z,yz^2,z^3$	$1+3T+5T^2+T^3$	$24T^{-1}+11+T$	1

No.	Generators	Hilbert series	Hilbert series of the normal module	exponential series	Hilbert
17.	$xy, x^3, x^2z, xz^2,$ y^3, y^2z, yz^2, z^3	$1+3T+5T^2$	$38T^{-1}+5$	1	
18.	$x^2, xy^2, xyz, xz^2,$ y^3, y^2z, yz^2, z^3	$1+3T+5T^2$	$38T^{-1}+5$	1	
19.	$x^2, yz, xy^2, xz^2,$ y^3, z^3	$1+3T+4T^2$	$22T^{-1}+8$	1	
20.	$x^2, y^2, xyz, xz^2,$ yz^2, z^3	$1+3T+4T^2$	$22T^{-1}+8$	1	
21.	x^2, xy, z^2, y^3, y^2z	$1+3T+3T^2$	$12T^{-1}+9$	1	
22.	x^2, y^2, z^2, xyz	$1+3T+3T^2$	$12T^{-1}+9$	1	
23.	x^2, xy, xz, y^2, z^2	$1+3T+T^2$	$10T^{-1}+5$	1	
24.	$x^2, xy, xz, y^2, yz,$ z^2	$1+3T$	$18T^{-1}$	1	
25.	x, y^3, y^2z, yz^2, z^3	$1+2T+3T^2$	$13T^{-1}+2+3T$	1	
26.	x, y^2, yz, z^2	$1+2T$	$7T^{-1}+2$	1	
27.	x, y, z	1	$3T^{-1}$	1	
28.	$x^3, x^2y, xyz, y^3,$ yz^2, z^3	$1+3T+6T^2+4T^3$ $+T^4$	$T^{-2}+18T^{-1}+20$ $+6T$	$1+T$	
29.	$x^3, x^2y, xyz, xz^2,$ y^3, z^3	$1+3T+6T^2+4T^3$ $+T^4$	$T^{-2}+18T^{-1}+20$ $+6T$	$1+T$	
30.	$x^3, x^2y, xy^2, xz^2,$ y^3, z^3	$1+3T+6T^2+4T^3$ $+T^4$	$T^{-2}+18T^{-1}+20$ $+6T$	$1+T$	
31.	$x^3, x^2y, x^2z, xyz,$ y^3, y^2z, z^3	$1\ +3T\ +6T^2$ $+3T^3$	$T^{-2}\ +25T^{-1}$ $+21$	$1+T$	
32.	$x^3, x^2y, x^2z, xy^2,$ y^3, yz^2, z^3	$1\ +3T\ +6T^2$ $+3T^3$	$T^{-2}\ +25T^{-1}$ $+21$	$1+T$	
33.	$x^3, x^2y, x^2z, y^3,$ y^2z, yz^2, z^3	$1\ +3T\ +6T^2$ $+3T^3$	$T^{-2}\ +25T^{-1}$ $+21$	$1+T$	
34.	$x^2, xyz, y^3, y^2z,$ z^3	$1\ +3T\ +5T^2$ $+3T^3$	$T^{-2}+15T^{-1}+17$ $+3T$	$1+T$	
35.	$x^2, y^3, y^2z, yz^2,$ z^3	$1\ +3T\ +5T^2$ $+3T^3$	$T^{-2}+15T^{-1}+17$ $+3T$	$1+T$	
36.	$x^2, xy^2, y^3, yz^2,$ z^3	$1\ +3T\ +5T^2$ $+3T^3$	$T^{-2}+15T^{-1}+17$ $+3T$	$1+T$	
37.	$xy, x^3, x^2z, xz^2,$ y^3, y^2z, z^3	$1+3T+5T^2+T^3$	$T^{-2}+25T^{-1}+11$ $+T$	$1+T$	
38.	$x^2, xy^2, xyz, xz^2,$ y^3, y^2z, z^3	$1+3T+5T^2+T^3$	$T^{-2}+25T^{-1}+11$ $+T$	$1+T$	
39.	$xy, x^3, x^2z, xz^2,$ y^3, yz^2, z^3	$1+3T+5T^2+T^3$	$T^{-2}+25T^{-1}+11$ $+T$	$1+T$	
40.	x^2, yz, xy^2, y^3, z^3	$1+3T+4T^2+T^3$	$T^{-2}+13T^{-1}+11$ $+2T$	$1+T$	
41.	$x^2, xy, y^3, y^2z,$ yz^2, z^3	$1+3T+4T^2+T^3$	$T^{-2}+13T^{-1}+11$ $+2T$	$1+T$	
42.	$x^2, y^2, xyz, xz^2,$ z^3	$1+3T+4T^2+T^3$	$T^{-2}+13T^{-1}+11$ $+2T$	$1+T$	

No.	Generators	Hilbert series	Hilbert series of the normal module	exponential Hilbert series
43.	$x^2, y^2, xz^2, yz^2, z^3$	$1+3T+4T^2+T^3$	$T^{-2}+13T^{-1}+11+2T$	$1+T$
44.	$x^2, xy, xz^2, y^3, y^2z, yz^2, z^3$	$1+3T+4T^2$	$T^{-2}+23T^{-1}+8$	$1+T$
45.	$xy, xz, x^3, y^3, y^2z, yz^2, z^3$	$1+3T+4T^2$	$T^{-2}+23T^{-1}+8$	$1+T$
46.	x^2, xy, y^2, z^2	$1+3T+2T^2$	$T^{-2}+9T^{-1}+8$	$1+T$
47.	x^2, xy, yz, z^2, y^3	$1+3T+2T^2$	$T^{-2}+9T^{-1}+8$	$1+T$
48.	x, y, z^2	$1+T$	$T^{-2}+3T^{-1}+2$	$1+T$
49.	$xy, x^3, x^2z, xz^2, y^3, z^3$	$1+3T+5T^2+2T^3+T^4$	$2T^{-2}+18T^{-1}+8+7T+T^2$	$1+2T+T^2$
50.	$x^2, xy^2, xyz, xz^2, y^3, z^3$	$1+3T+5T^2+2T^3+T^4$	$2T^{-2}+18T^{-1}+8+7T+T^2$	$1+2T+T^2$
51.	x, yz, y^3, z^3	$1+2T+2T^2$	$2T^{-2}+7T^{-1}+4+2T$	$1+2T+T^2$
52.	x, y^2, yz^2, z^3	$1+2T+2T^2$	$2T^{-2}+7T^{-1}+4+2T$	$1+2T+T^2$
53.	x, y^2, yz, z^3	$1+2T+T^2$	$2T^{-2}+5T^{-1}+4+T$	$1+2T+T^2$
54.	x, y^2, z^2	$1+2T+T^2$	$2T^{-2}+5T^{-1}+4+T$	$1+2T+T^2$
55.	x, y, z^3	$1+T+T^2$	$T^{-3}+T^{-2}+3T^{-1}+2+2T$	$1+2T+2T^2+T^3$
56.	x^3, xyz, y^3, z^3	$1+3T+6T^2+6T^3+3T^4$	$3T^{-2}+18T^{-1}+24+12T$	$1+3T+3T^2+T^3$
57.	$x^3, x^2y, y^3, yz^2, z^3$	$1+3T+6T^2+5T^3+2T^4$	$3T^{-2}+17T^{-1}+21+10T$	$1+3T+3T^2+T^3$
58.	x^3, x^2y, xyz, y^3, z^3	$1+3T+6T^2+5T^3+2T^4$	$3T^{-2}+17T^{-1}+21+10T$	$1+3T+3T^2+T^3$
59.	$x^3, x^2y, xz^2, y^3, z^3$	$1+3T+6T^2+5T^3+2T^4$	$3T^{-2}+17T^{-1}+21+10T$	$1+3T+3T^2+T^3$
60.	$x^3, x^2y, xy^2, xyz, y^3, z^3$	$1+3T+6T^2+4T^3+2T^4$	$3T^{-2}+19T^{-1}+14+12T$	$1+3T+3T^2+T^3$
61.	$x^3, x^2y, x^2z, xy^2, y^3, z^3$	$1+3T+6T^2+4T^3+2T^4$	$3T^{-2}+19T^{-1}+14+12T$	$1+3T+3T^2+T^3$
62.	$x^3, x^2y, x^2z, xy^2, y^3, y^2z, z^3$	$1+3T+6T^2+3T^3+T^4$	$3T^{-2}+24T^{-1}+14+7T$	$1+3T+3T^2+T^3$
63.	$x^3, x^2y, x^2z, xy^2, xyz, y^3, z^3$	$1+3T+6T^2+3T^3+T^4$	$3T^{-2}+24T^{-1}+14+7T$	$1+3T+3T^2+T^3$
64.	$x^3, x^2y, x^2z, xy^2, xyz, y^3, y^2z, z^3$	$1+3T+6T^2+2T^3$	$3T^{-2}+33T^{-1}+16$	$1+3T+3T^2+T^3$
65.	$xy, x^3, xz^2, y^3, yz^2, z^3$	$1+3T+5T^2+2T^3$	$3T^{-2}+19T^{-1}+15+2T$	$1+3T+3T^2+T^3$
66.	$xy, x^3, x^2z, y^3, yz^2, z^3$	$1+3T+5T^2+2T^3$	$3T^{-2}+19T^{-1}+15+2T$	$1+3T+3T^2+T^3$
67.	$x^2, xy^2, xyz, y^3, y^2z, z^3$	$1+3T+5T^2+2T^3$	$3T^{-2}+19T^{-1}+15+2T$	$1+3T+3T^2+T^3$

No.	Generators	Hilbert series	Hilbert series of the normal module	exponential Hilbert series
68.	$xy,\, x^3,\, x^2z,\, y^3,\, y^2z, z^3$	$1\ +3T\ +5T^2\ +2T^3$	$3T^{-2}\ +19T^{-1}\ +15+2T$	$1+3T+3T^2+T^3$
69.	$x^2,\, xy^2,\, xz^2,\, y^3,\, y^2z, z^3$	$1\ +3T\ +5T^2\ +2T^3$	$3T^{-2}\ +19T^{-1}\ +15+2T$	$1+3T+3T^2+T^3$
70.	x^2, y^2, xyz, z^3	$1\ +3T\ +4T^2\ +2T^3$	$3T^{-2}\ +11T^{-1}\ +12+4T$	$1+3T+3T^2+T^3$
71.	x^2, xy, y^3, yz^2, z^3	$1\ +3T\ +4T^2\ +2T^3$	$3T^{-2}\ +11T^{-1}\ +12+4T$	$1+3T+3T^2+T^3$
72.	x^2, yz, y^3, z^3	$1\ +3T\ +4T^2\ +2T^3$	$3T^{-2}\ +11T^{-1}\ +12+4T$	$1+3T+3T^2+T^3$
73.	x^2, xy, y^3, y^2z, z^3	$1\ +3T\ +4T^2\ +2T^3$	$3T^{-2}\ +11T^{-1}\ +12+4T$	$1+3T+3T^2+T^3$
74.	x^2, y^2, xz^2, z^3	$1\ +3T\ +4T^2\ +2T^3$	$3T^{-2}\ +11T^{-1}\ +12+4T$	$1+3T+3T^2+T^3$
75.	x^2, xy, y^2, xz^2, z^3	$1+3T+3T^2+T^3$	$3T^{-2}\ +9T^{-1}\ +9 +3T$	$1+3T+3T^2+T^3$
76.	x^2, y^2, z^2	$1+3T+3T^2+T^3$	$3T^{-2}\ +9T^{-1}\ +9 +3T$	$1+3T+3T^2+T^3$
77.	x^2, xy, z^2, y^3	$1+3T+3T^2+T^3$	$3T^{-2}\ +9T^{-1}\ +9 +3T$	$1+3T+3T^2+T^3$
78.	x^2, xy, yz, y^3, z^3	$1+3T+3T^2+T^3$	$3T^{-2}\ +9T^{-1}\ +9 +3T$	$1+3T+3T^2+T^3$
79.	$x^2, xy, yz, xz^2, y^3, z^3$	$1+3T+3T^2$	$3T^{-2}\ +15T^{-1}\ +9$	$1+3T+3T^2+T^3$
80.	$xy, xz, yz, x^3, y^3, z^3$	$1+3T+3T^2$	$3T^{-2}\ +15T^{-1}\ +9$	$1+3T+3T^2+T^3$
81.	$x^2,\, xy,\, y^2,\, xz^2, yz^2, z^3$	$1+3T+3T^2$	$3T^{-2}\ +15T^{-1}\ +9$	$1+3T+3T^2+T^3$
82.	x^2, xy, xz, y^2, z^3	$1+3T+2T^2+T^3$	$3T^{-2}\ +9T^{-1}\ +5 +4T$	$1+3T+3T^2+T^3$
83.	$x^2, xy, xz, y^2, yz, z^3$	$1+3T+T^2$	$3T^{-2}\ +13T^{-1}\ +5$	$1+3T+3T^2+T^3$
84.	x, y^3, y^2z, z^3	$1+2T+3T^2+T^3$	$3T^{-2}\ +9T^{-1}\ +5 +3T+T^2$	$1+3T+3T^2+T^3$
85.	$x^3, x^2y, x^2z, xyz, y^3, z^3$	$1+3T+6T^2+4T^3 +T^4$	$4T^{-2}\ +21T^{-1}\ +20+6T$	$1+4T+6T^2+4T^3+T^4$
86.	$x^3, x^2y, x^2z, y^3, y^2z, z^3$	$1+3T+6T^2+4T^3 +T^4$	$4T^{-2}\ +21T^{-1}\ +20+6T$	$1+4T+6T^2+4T^3+T^4$
87.	x^2, y^3, y^2z, z^3	$1+3T+5T^2+4T^3 +T^4$	$4T^{-2}\ +14T^{-1}\ +16+7T+T^2$	$1+4T+6T^2+4T^3+T^4$
88.	x^2, xyz, y^3, z^3	$1+3T+5T^2+4T^3 +T^4$	$4T^{-2}\ +14T^{-1}\ +16+7T+T^2$	$1+4T+6T^2+4T^3+T^4$
89.	xy, x^3, x^2z, y^3, z^3	$1+3T+5T^2+3T^3 +T^4$	$4T^{-2}\ +14T^{-1}\ +13+7T+T^2$	$1+4T+6T^2+4T^3+T^4$
90.	x^2, xy^2, xyz, y^3, z^3	$1+3T+5T^2+3T^3 +T^4$	$4T^{-2}\ +14T^{-1}\ +13+7T+T^2$	$1+4T+6T^2+4T^3+T^4$
91.	$x^2, xy^2, y^3, y^2z, z^3$	$1+3T+5T^2+3T^3 +T^4$	$4T^{-2}\ +14T^{-1}\ +13+7T+T^2$	$1+4T+6T^2+4T^3+T^4$

No.	Generators	Hilbert series	Hilbert series of the normal module	exponential Hilbert series
92.	xy, x^3, xz^2, y^3, z^3	$1+3T+5T^2+3T^3 +T^4$	$4T^{-2} +14T^{-1} +13+7T+T^2$	$1+4T+6T^2+4T^3+T^4$
93.	$x^2, xy^2, xz^2, y^3, z^3$	$1+3T+5T^2+3T^3 +T^4$	$4T^{-2} +14T^{-1} +13+7T+T^2$	$1+4T+6T^2+4T^3+T^4$
94.	$x^2, xy, xz^2, y^3, yz^2, z^3$	$1+3T+4T^2+T^3$	$4T^{-2} +16T^{-1} +11+2T$	$1+4T+6T^2+4T^3+T^4$
95.	$x^2, xy, xz^2, y^3, y^2z, z^3$	$1+3T+4T^2+T^3$	$4T^{-2} +16T^{-1} +11+2T$	$1+4T+6T^2+4T^3+T^4$
96.	$xy, xz, x^3, y^3, y^2z, z^3$	$1+3T+4T^2+T^3$	$4T^{-2} +16T^{-1} +11+2T$	$1+4T+6T^2+4T^3+T^4$
97.	$x^2, xy, xz, y^3, y^2z, yz^2, z^3$	$1+3T+3T^2$	$4T^{-2} +16T^{-1} +9$	$1+4T+6T^2+4T^3+T^4$
98.	$x^2, xy, xz, y^2, yz^2, z^3$	$1+3T+2T^2$	$4T^{-2} +12T^{-1} +8$	$1+4T+6T^2+4T^3+T^4$
99.	$x^2, xy, xz, yz, y^3, z^3$	$1+3T+2T^2$	$4T^{-2} +12T^{-1} +8$	$1+4T+6T^2+4T^3+T^4$
100.	$x^3, x^2y, xy^2, y^3, z^3$	$1+3T+6T^2+5T^3 +3T^4$	$T^{-3} +3T^{-2} +18T^{-1} +17 +15T$	$1 +4T +7T^2 +7T^3 +4T^4 +T^5$
101.	x^2, xy, y^2, z^3	$1 +3T +3T^2 +2T^3$	$T^{-3} +3T^{-2} +9T^{-1}+8+6T$	$1 +4T +7T^2 +7T^3 +4T^4 +T^5$
102.	x, y^2, z^3	$1+2T+2T^2+T^3$	$T^{-3} +3T^{-2} +5T^{-1} +5 +3T +T^2$	$1 +4T +7T^2 +7T^3 +4T^4 +T^5$
103.	x^2, xy, xz^2, y^3, z^3	$1+3T+4T^2+2T^3 +T^4$	$5T^{-2}+12T^{-1}+8 +6T+2T^2$	$1+5T+10T^2+10T^3+5T^4 +T^5$
104.	xy, xz, x^3, y^3, z^3	$1+3T+4T^2+2T^3 +T^4$	$5T^{-2}+12T^{-1}+8 +6T+2T^2$	$1+5T+10T^2+10T^3+5T^4 +T^5$
105.	x^2, xy, xz, y^3, z^3	$1+3T+3T^2+2T^3 +T^4$	$6T^{-2}+10T^{-1}+6 +5T+3T^2$	$1 +6T +15T^2 +20T^3 +15T^4+6T^5+T^6$
106.	$x^2, xy, xz, y^3, y^2z, z^3$	$1+3T+3T^2+T^3$	$6T^{-2}+12T^{-1}+9 +3T$	$1 +6T +15T^2 +20T^3 +15T^4+6T^5+T^6$
107.	x^2, xy^2, y^3, z^3	$1+3T+5T^2+4T^3 +2T^4$	$T^{-3} +5T^{-2} +13T^{-1} +14 +10T+2T^2$	$1 +6T +16T^2 +25T^3 +25T^4+16T^5+6T^6+T^7$
108.	xy, x^3, y^3, z^3	$1+3T+5T^2+4T^3 +2T^4$	$T^{-3} +5T^{-2} +13T^{-1} +14 +10T+2T^2$	$1 +6T +16T^2 +25T^3 +25T^4+16T^5+6T^6+T^7$
109.	x^2, y^2, z^3	$1+3T+4T^2+3T^3 +T^4$	$T^{-3} +5T^{-2} +10T^{-1}+11+7T +2T^2$	$1 +6T +16T^2 +25T^3 +25T^4+16T^5+6T^6+T^7$
110.	x^2, xy, y^3, z^3	$1+3T+4T^2+3T^3 +T^4$	$T^{-3} +5T^{-2} +10T^{-1}+11+7T +2T^2$	$1 +6T +16T^2 +25T^3 +25T^4+16T^5+6T^6+T^7$
111.	x, y^3, z^3	$1+2T+3T^2+2T^3 +T^4$	$2T^{-3} +4T^{-2} +7T^{-1} +6 +5T +2T^2+T^3$	$1 +6T +17T^2 +30T^3 +36T^4 +30T^5 +17T^6 +6T^7+T^8$

No.	Generators	Hilbert series	Hilbert series of the normal module	exponential Hilbert series
112.	$x^3, x^2y, x^2z, y^3,$ z^3	$1+3T+6T^2+5T^3$ $+3T^4+T^5$	$7T^{-2}+18T^{-1}$ $+15+12T+5T^2$	$1+7T+21T^2+35T^3$ $+35T^4+21T^5+7T^6+T^7$
113.	x^3, x^2y, y^3, z^3	$1+3T+6T^2+6T^3$ $+4T^4+T^5$	$T^{-3}+6T^{-2}$ $+17T^{-1}+20$ $+15T+4T^2$	$1+7T+22T^2+41T^3$ $+50T^4+41T^5+22T^6$ $+7T^7+T^8$
114.	x^2, y^3, z^3	$1+3T+5T^2+5T^3$ $+3T^4+T^5$	$2T^{-3}+7T^{-2}$ $+13T^{-1}+15$ $+11T+5T^2+T^3$	$1+9T+38T^2+100T^3$ $+183T^4+245T^5+245T^6$ $+183T^7+100T^8+38T^9$ $+9T^{10}+T^{11}$
115.	x^3, y^3, z^3	$1+3T+6T^2+7T^3$ $+6T^4+3T^5+T^6$	$3T^{-3}+9T^{-2}$ $+18T^{-1}+21$ $+18T+9T^2$ $+3T^3$	$1+12T+69T^2+253T^3$ $+663T^4+1317T^5$ $+2050T^6+2547T^7$ $+2547T^8+2050T^9$ $+1317T^{10}+663T^{11}$ $+253T^{12}+69T^{13}+12T^{14}$ $+T^{15}$

9.5. Germs in 4-space with tangent cone defined in degree 2

Torsten Ekedahl and Jan-Erik Roos (Stockholm Math Institute) wrote a script for the Bayer-Stillman computer algebra system Macaulay calculating the generator degrees of the normal modules for all homogeneous singularities of embedding dimension 4 defined by quadratic equations. It turned out that there are only 32 different generator degree types. With the exception of one case, all generator types give singularities such that all deformations are tangentially flat. In the exceptional case there is just one generator of degree -2 in the normal module and no further element in lower degrees. The argument of 9.3 gives an inequality

$$\mathrm{H}_A^1 \cdot (1 + 4T) \le \mathrm{H}_B^1 \cdot (1 + T),$$

i.e., $\mathrm{H}_A^1(d) + 4\,\mathrm{H}_A^1(d-1) \le \mathrm{H}_B^1(d) + \mathrm{H}_B^1(d-1)$ for every d. In particular, the multiplicities satisfy $5e_0(A) \le 2e_0(B)$. On the other hand, multiplication by $(1-T^2)^{-1}$ of the above inequality gives $\mathrm{H}_A^2 \le \mathrm{H}_B^2$. In fact this phenomenon can be generalized to arbitrary dimension using 9.2 and a careful estimation of the numbers $n(d)$ (see [Ri93]).

Corollary

Let $f : (A, m, K) \to (B, n, L)$ be a residually separable flat local homomorphism of local rings such that $B/mB \cong K[x, y, z, w]/I$ with the initial ideal $\mathrm{in}(I)$ of I

generated in degree two. Then

$$H_A^2 \leq H_B^2$$

Here K is the finite field with 31991 elements.

We conclude the chapter formulating several questions and open problems.

9.6. Problem: generalization of the main theorem

Can the main theorem be proved without the assumption that the homomorphism is residually separable. Moreover, is it possible to weaken the assumption that the involved rings are equipped with the natural filtrations, i.e., can one prove

$$H_A^1 \cdot H_{B/mB}^0 \leq H_B^1 \cdot \prod_{d=2}^{\infty} \left(\frac{1 - T^d}{1 - T} \right)^{n(d)}$$

for more general flat homomorphisms $f : (A, m, K) \to (B, n, L)$ of filtered local rings?

9.7. Problem: injectivity of the Kodaira-Spencer map

Let $f : (A, m, K) \to (B, n, L)$ be a residually purely inseparable flat local homomorphism of filtered local rings. Is it possible to prove in this situation that the Kodaira-Spencer map ϕ of 8.13 is itself injective in the indicated degrees? For this it would be sufficient to show that ϕ is definable over K. We can prove the latter only for a modified Kodaira-Spencer map involving a Frobenius action. Unfortunately, we don't know whether the analog of 8.13 holds for this modified Kodaira-Spencer map. If it would, the restriction to residually separable homomorphisms in the main theorem could be skipped.

9.8. Problem: are there further inequalities?

Do the formulas

$$H_B^1 \leq H_A^1 \cdot H_{B/mB}^0$$

and

$$H_A^1 \cdot H_{B/mB}^0 \leq H_B^1 \cdot \prod_{d=2}^{\infty} \left(\frac{1 - T^d}{1 - T} \right)^{n(d)}$$

represent the first two members of an infinite sequence of estimations giving step by step a closer relation between H_A^1 and H_B^1?

9.9. Problem: estimation of the product on the right

Given a flat local homomorphism $f : (A, m, K) \to (B, n, L)$ find an estimation of the product

$$\prod_{d=2}^{\infty} \left(\frac{1 - T^d}{1 - T} \right)^{n(d)}$$

associated with Schlessinger's T^1 of $G(B/mB)$ in terms of $L(B/mB)$.

9.10. Problem: Cohen factorization

Factors every flat homomorphism of cofinitely filtered complete local rings into a tangentially flat one and a surjection? In other words, to which extent can 5.9 be generalized?

9.11. Problem: non-flat extensions

Prove formulas relating H_A^1 and H_B^1 for local homomorphisms $(A, m, K) \to (B, n, L)$ of local rings that are not necessarily flat. A variant of the *Russian Conjecture* (see Popov's Berkeley talk [Po]) can be formulated in this context.

9.12. Problem: $F(f)$ under composition

Find descriptions of the minimal flatifying filtrations for a given homomorphism of filtered local rings that allow to study its behavior under extension of the residue class field. Given a composition $A \to B \to C$ of flat homomorphisms of filtered local rings such that the maximal ideal of A generates the maximal ideal of B, what are the relations between the flatifying filtrations of $A \to C$ and $B \to C$?

9.13. Problem: Hilbert series for different filtrations

Consider the ideal $I := (xy, xz, yz, y^2, z^2, x^3)$ of the polynomial ring $R := K[x, y, z]$. This is one of the "bad" examples from Table 9.4 above (see example 83 of 9.4 and example (2.27) of [Ja90]). Schlessinger's T^1 of the factor ring R/I has three generators of degree -2 (relative to the natural filtrations). Now let $q := (\frac{1}{3}, \frac{2}{3}, \frac{2}{3})$ and equip R with the q-adic filtration. A Macaulay computation shows that the graded ring $G(R/I)$ has a normal module without non-zero homogeneous elements of degrees less than -1. This can be used to prove Lech's inequality on the multiplicity level for flat local homomorphisms with special fibre $K[x, y, z]/I$.

Similar results can be obtained using other "bad" singularities from the above table. As shown by J. Jahnel, this is a general phenomenon (see [Ja90], Proposition (3.26) and its proof). Given a local ring B_0 there is always a filtration of B_0 even an ideal-adic one, such that Schlessinger's T^1 of the associated graded ring doesn't have any non-zero homogeneous elements of degrees less than -1. This implies that all deformations $A \rightarrow B$ of B_0 with respect to this filtration are tangentially flat (see [Ja90] for terminology). As a consequence one obtains relations between weighted versions of the Hilbert-Samuel series H_A^1 and H_B^1 (i.e., Hilbert-Samuel series associated with non-natural filtrations). Such relations often imply Lech's inequality for the usual Hilbert-Samuel series. So it is natural to ask whether there is always a filtration suited to prove Lech's inequality, or more generally to ask for a description of the set of filtrations of B_0 such that T^1 of $G(B_0)$ doesn't have any non-zero homogeneous elements of degrees less than -1.

10. On the local rings of the Hilbert scheme

Remark

This section is intended as an extended illustration of the previously introduced concepts. We want to consider a given point on the Hilbert scheme corresponding to a fixed singularity $\bar{B}$ with respect to some (formal local) embedding into a regular space. Let $f : A \rightarrow B$ be the germ of the universal family at that point. Consider it as a homomorphism of filtered local rings equipped with the natural filtrations. Our aim is to give a description of the minimal flatifying filtration of f. As a side result we will get an alternative proof for the tangential flatness criterion [He92, Th. 2.5] of deformations in terms of the normal module (the original proof is very technical). Moreover, we clarify the situation when this criterion cannot be applied. It turns out that there are only finitely many obstructions for a deformation to be tangentially flat.

All considerations will be restricted to the case that $\bar{B}$ is an Artinian local ring corresponding to a single K-rational point in the total space of the universal family, though everything can be done also quite generally. In the special case we can avoid technical difficulties and considerably simplify the presentation. We briefly recall the concepts related with the notion of Hilbert scheme.

10.1. Hilbert scheme point functor

Let S be a locally Noetherian scheme and $X \rightarrow S$ a projective scheme over S, i.e., X is a closed subscheme of the projective bundle $\mathbb{P}(E)$ where E is a locally free O_S-module of finite rank. Then the functor

$$\text{Hilb}_{X/S} : \quad S\text{-schemes} \longrightarrow \text{sets}$$

such that

$$\text{Hilb}_{X/S}(S') := \{Y \subseteq X \times_S S' | Y \text{ a closed subscheme flat over } S'\}$$

is called the *Hilbert scheme point functor* of X/S.

10.2. Representability

The functor $\text{Hilb}_{X/S} : S\text{-schemes} \longrightarrow$ sets is representable by a locally Noetherian S-scheme $\text{Hilb}(X/S)$, the *Hilbert scheme* of X over S.

For a proof see [Gr60].

10.3. A series of subfunctors

Let $p \in \mathbb{Q}[T]$ be a polynomial over the rational numbers. Then the subfunctor of $\mathrm{Hilb}_{X/S}$ defined by

$$\mathrm{Hilb}^p_{X/S}(S') := \{Y \in \mathrm{Hilb}_{X/S}(S')|\ Y/S \text{ has fibres with Hilbert polynomial } p\}$$

is called the Hilbert scheme point functor associated with p.

10.4. Decomposition of $\mathrm{Hilb}(X/S)$

(ii) $\mathrm{Hilb}(X/S) = \cup_{p \in \mathbb{Q}[T]} \mathrm{Hilb}^p(X/S)$ (disjoint union)

(iii) If S is connected, then so is $\mathrm{Hilb}^p(X/S)$.

See [Mu66], [D-C70], [Gr60], [Ha66], for details and [Do72] for hints to further literature.

10.5. Terminology

In that what follows we will use the above concepts in the following special situation.

$S := \mathrm{Spec}\, K$,	K a field (will be replaced later by a finite extension to get K-rational points)
$X := \mathbb{P}^N_K$	projective N-space over K
$p(T) := \nu$	a constant polynomial
$\mathbb{H} := \mathrm{Hilb}^n(\mathbb{P}^N_K/K)$	the representing object of $\mathrm{Hilb}^n_{\mathbb{P}^N_K/K}$
$\phi : \mathbb{U} \subseteq \mathbb{P}^N_{\mathbb{H}} := \mathbb{P}^N_K \times_K \mathbb{H} \to \mathbb{H}$	the universal family
$0 \in \mathbb{H}$	a fixed point corresponding to our given singularity $\bar{B}$ ($=$ special fibre of $f : A \to B$). For simplicity we will assume that 0 is a K-rational point of $\mathbb{H}$.
$\mathbb{U}_0 := \phi^{-1}(0)$	the (scheme theoretic) fibre of ϕ over $0 \in \mathbb{H}$
$A := O_{\mathbb{H},0}$	the local ring of $\mathbb{H}$ at 0
$m := m(A)$	the maximal ideal of A
$B := O_{\mathbb{U},\mathbb{U}_0}$	the germs of regular functions on $\mathbb{U}$ along the special fibre (we will restrict to the case that $\mathbb{U}_0 = \{0\}$ consists of a single point so that B is a local ring)
$n := m(B)$	the maximal ideal of B
$f : A \to B$	the local homomorphism induced by ϕ
$\bar{B} := B/mB$	the germs of regular functions on the special fibre

144

In that what follows, an *embedded deformation germ* will be a sequence of ring homomorphisms

$$f' : A' \xrightarrow{i'} A'[X_1, \ldots, X_N] \xrightarrow{j'} B'$$

where (A', m') is a local K-algebra with K mapping surjectively to the residue class field (i.e., $A'/m' = K$), i' is the natural embedding into the polynomial extension with N being the dimension of the projective space $\mathbb{P}_K^N$ above, and j' is a surjection such that B' is a free module of rank ν over A', ν being the only value of the constant polynomial $p(T)$ above. Later on we will additionally require that the ring B' is also a local K-algebra with residue class field K, in which case the composition $f' = j' \circ i'$ is automatically a local homomorphism (since B' is free, hence faithfully flat over A'). A *homomorphism of embedded deformation germs* is defined to be a commutative diagram of ring homomorphisms

$$
\begin{array}{ccccc}
A' & \xrightarrow{i'} & A'[X_1, \ldots, X_N] & \xrightarrow{j'} & B' \\
\downarrow & & \downarrow & & \downarrow \\
A'' & \xrightarrow{i''} & A''[X_1, \ldots, X_N] & \xrightarrow{j''} & B''
\end{array} ,
$$

where the two rows are embedded deformation germs, the left hand side vertical homomorphism is local and the one in the middle induces the identity map on the set of indeterminates. If f' and f'' denote, respectively, the upper and the lower row of this diagram, we will sometimes write $f' \to f''$ to denote this homomorphism.

Let f' be an embedded deformation germ as above and $g : A' \to A''$ a residually rational local K-algebra homomorphism. Then $f' \otimes_{A'} A''$ is again an embedded deformation germ. Below $f' \otimes_{A'} A''$ will be called the embedded deformation obtained from f' by *base change* with respect to the *base change homomorphism* $g : A' \to A''$.

Assume that $f' : A \to A'[X_1, \ldots, X_N] \to B'$ is an embedded deformation germ such that B' is local. Then this property is preserved under base change. For, if $g : A' \to A''$ is the (residually rational) base change homomorphism, the ring $B' \otimes_{A'} A''$ is local, since it is finite over A'' and has a special fibre which is a local ring (the same like the special fibre of B' over A').

Now assume that the rings B' and B'' in the above homomorphism $f' \to f''$ of embedded deformation germs are local with residue class field K. Then the right hand side vertical homomorphism $B' \to B''$ is automatically local. In order to see this, consider the factorization $f' \to f' \otimes_{A'} A'' \to f''$. The first morphism $f' \to f' \otimes_{A'} A''$ has the required property (since $B' \otimes_{A'} A''$ is local as we just have seen, and $A' \to A''$ is local by assumption), hence it suffices to prove the claim for the second homomorphism $f' \otimes_{A'} A'' \to f''$. This reduces us to the case $A' = A''$ and $A' \to A''$ equal to the identity map. Since composition of the horizontal maps in the above commutative diagram gives local homomorphisms, we may tensor the whole diagram with A'/m' without loosing any maximal ideal, i.e., we may assume $A'' = A' = K$. But then $B' \to B''$ must be local, since it is a homomorphism of Artinian local algebras (so that the maximal ideals are nilpotent).

Obviously, embedded deformation germs together with the above defined morphisms form a category. In particular, we can speak about isomorphic embedded deformation germs and we can form isomorphism classes. Isomorphic embedded deformation germs will be treated below essentially as if they were equal.

10.6. The universality property of $\mathbb{U} \to \mathbb{H}$

The universal family ϕ is characterized by the property that it is flat and that for every flat morphism ϕ' that can be decomposed

$$f' : \mathbb{U}' \subseteq \mathbb{P}_K^N \times_K \mathbb{H}' \to \mathbb{H}'$$

into a closed embedding and the natural projection there is a unique base change morphism

$$\eta : \mathbb{H}' \to \mathbb{H}$$

inducing a morphism $\mathbb{U}' \to \mathbb{U}$ such that the diagram

$$
\begin{array}{ccccc}
\mathbb{U} & \subseteq & \mathbb{P}_K^N \times_K \mathbb{H} & \to & \mathbb{H} \\
\uparrow & & \uparrow & & \uparrow \eta \\
\mathbb{U}' & \subseteq & \mathbb{P}_K^N \times_K \mathbb{H}' & \to & \mathbb{H}'
\end{array}
$$

is (commutative and) cartesian.

10.7. The germ of the universal family at 0

Let

$$f : B := O_{\mathbb{U}, \mathbb{U}_o} \leftarrow A := O_{\mathbb{H}, 0}$$

be the ring homomorphism induced by the universal family $\phi : \mathbb{U} \subseteq \mathbb{P}_{\mathbb{H}}^N := \mathbb{P}_K^N \times_K \mathbb{H} \to \mathbb{H}$. Then f factors

$$f : A \xrightarrow{i} A[X_1, \ldots, X_N] \xrightarrow{j} B$$

into a polynomial extension i and a surjection j giving this way an embedded deformation germ (see 10.5) such that the germ obtained by base change with respect to the natural surjection $A \to A/m = K$ is just the embedded singularity $f_0 : K \to K[X_1, \ldots, X_N] \to \bar{B}$ defining the point $0 \in \mathbb{H}$. Moreover, an arbitrary embedded deformation germ

$$f' : A' \xrightarrow{i'} A'[X_1, \ldots, X_N] \xrightarrow{j'} B'$$

such that $f' \otimes_{A'} A'/m' = f_0$ can be obtained from f as a result of base change with respect to an uniquely determined base change homomorphism $g : A \to A'$. Below we will refer to $f : A \to A[X] \to B$ as to the *universal germ*.

Proof. By construction, the universal family

$$\phi : \mathbb{U} \subseteq \mathbb{P}_{\mathbb{H}}^N := \mathbb{P}_K^N \times_K \mathbb{H} \to \mathbb{H}$$

is a projective morphism, hence of finite type. Since the Hilbert polynomial of the fibre is assumed to be a constant, the morphism ϕ is quasi-finite. Zariski's Main theorem applied to ϕ yields a decomposition

$$\mathbb{U} \subseteq \mathbb{U}' \to \mathbb{H}$$

of ϕ into an open embedding $\mathbb{U} \subseteq \mathbb{U}'$ and a finite morphism $\mathbb{U}' \to \mathbb{H}$. Since $\mathbb{U}$ is projective over $\mathbb{H}$, the open embedding $\mathbb{U} \subseteq \mathbb{U}'$ is simultaneously a closed embedding, i.e., $\phi : \mathbb{U} \to \mathbb{H}$ itself is already finite. In particular, locally around a given point of $\mathbb{H}$, the projective space $\mathbb{P}_K^N$ can be replaced by some affine subspace. So there are affine neighborhoods U of $0 \in \mathbb{H}$ and V of $\phi^{-1}(0) \subseteq \mathbb{U}$, such that ϕ induces a finite flat homomorphism

$$\phi|_V : V \subseteq \mathbb{A}_K^N \times_K U \to U$$

Note that the restriction $\phi|_V$ has a universality property similar to that of ϕ: given a morphism $\phi' : \mathbb{U}' \subseteq \mathbb{A}_K^N \times_K \mathbb{H}' \to \mathbb{H}'$ and a point $0' \in \mathbb{H}'$ such that the fibre over $0'$ (as a subscheme of $\mathbb{A}_K^N$) is equal to the fibre $\phi^{-1}(0)$ one can restrict the morphism ϕ' to an appropriate neighborhood of $0'$ such that there is a cartesian diagram as above with ϕ replaced by $\phi|_V$. Taking the homomorphism induced on the germs of regular functions along the fibre $\mathbb{U}_0$ over $0 \in \mathbb{H}$ we get the required characterization in terms of local rings.

10.8. Conventions

From the above we see that the special fibre $\bar{B} = B/mB$ is an Artinian ring, hence semi-local. Using the Chinese remainder theorem one easily sees that $\bar{B}$ is the direct product of Artinian local rings. In that what follows we will even assume that $\bar{B}$ is a local K-algebra with residue class field K. Since B is finite over A (i.e., each maximal ideal of B restricts to a maximal ideal of A), this implies that B is a local ring. As usual, assume that

$$n := m(B) \text{ is the maximal ideal of } B.$$

After a linear change of coordinates one may assume that the indeterminates $X_1, \ldots, X_N$ are mapped to elements of the maximal ideal n of B. This implies that the residue classes of the X_i's in B are nilpotent. So the rings B and $\bar{B}$ can be considered as factor rings of a power series ring over, respectively, A and K,

$$\left. \begin{array}{l} B = A[[X]]/I \\ \bar{B} = K[[X]]/\bar{I} \end{array} \right\} X := (X_1, \ldots, X_N).$$

Since $\bar{B}$ is Artinian and B is finite over A, the power series rings could of course be replaced by polynomial rings, but power series have the advantage that we remain inside the category of local rings. So we will sometimes prefer the power series notation.

In that what follows the indeterminate X_i and its residue classes in B and $\bar{B}$, respectively, will be usually denoted by one and the same symbol X_i. It will be always clear from the context which of the possible three elements we have in mind.

10.9. Standard monomials

Equip the set of power products in the variables $X_1, \ldots, X_N$,

$$P := \{X^i = X_1^{i_1} \cdot \ldots \cdot X_N^{i_N} \mid i = (i_1, \ldots, i_N) \in \mathbb{N}^N\},$$

with the *degreewise lexicographic order*:

$$X^i \geq X^j \quad \Leftrightarrow \quad |i| \leq |j| \text{ and, in case } |i| = |j|, \text{ the first non-zero}$$
coordinate of the difference $i - j$ is positive.

Here as usual, $|i| = i_1 + \ldots + i_N$ if $i = (i_1, \ldots, i_N)$. The *initial form* of a non-zero series $f = \sum c_i X^i \in K[X]$ is by definition the monomial

$$\mathrm{in}(f) = c_{i_0} X^{i_0}$$

such that $c_{i_0} \neq 0$ and $X^{i_0} \leq X^i$ whenever $c_i \neq 0$. A *standard monomial* (for $\bar{B} := K[[X_1, \ldots, X_N]]/\bar{I}$) is a power product X^i which is not an initial form of any element of $\bar{I}$,

$$X^i \neq \mathrm{in}(f) \text{ for every } f \in \bar{I}.$$

The set of standard monomials will be denoted by

$$S := \{X^i \in P \mid X^i \text{ a standard monomial}\}.$$

Further let

$$S' := \{X_i \in P \mid X_i \notin S\}$$

denote the set of non-standard variables. By definition, for each element $X_i \in S'$ there is a generator of $\bar{I}$ with initial form X_i.

10.10. Properties of the standard monomials

(i) $X^i | X^j$ and $X^j \in S$ implies $X^i \in S$.

(ii) If $X^i \in P - S$ then, as an element of $\bar{B}$, the monomial X^i can be written as a linear combination of lexicographically later standard monomials, i.e.,

$$X^i = \sum_{X^j \in S} c_{ij} X^j$$

with coefficients c_{ij} from K such that $c_{ij} = 0$ for $X^i \geq X^j$.

(iii) S is a vector space basis of $\bar{B}$ over K.

(iv) S is a free system of generators of B over A.

Proof. (i). There is some power product α with $X^j = \alpha \cdot X^i$. If X^i is not in S, then $X^i = \text{in}(f)$ for some $f \in \bar{I}$, hence $X^j = \text{in}(\alpha \cdot f)$ so that $X^j \notin S$, a contradiction.

(ii). Write

$$n(X^j) = k$$

if X^j is the k-th power product in the lexicographically ordered sequence of all power products. For example $n(1) = 1$, $n(X_1) = 2$, $n(X_2) = 3$, etc. Since $(X_1, \ldots, X_N)\bar{B}$ is nilpotent, all power products of large degrees are zero in $\bar{B}$. So there is some k_0 with

$$X^i = 0 \text{ in } \bar{B} \text{ provided } n(X^i) \geq k_0.$$

In particular, the power products X^i with $n(X^i) \geq k_0$ can trivially be written in the required form (with all coefficients c_{ij} equal to zero).

The further proof precedes by descending induction on $n(X^i)$. So assume that, $n(X^i) = k$, $X^i \notin S$ and that all power products $X^{i'}$ not in S with $n(X^{i'}) > k$ can be written in the required form. Since X^i is not in S, $X^i = \text{in}(f)$ for some $f \in \bar{I}$, hence in $\bar{B}$,

$$X^i = \sum_{X^i < X^j \in P} c_{ij} X^j, \quad c_{ij} \in K,$$

i.e., X^i is a K-linear combination of lexicographically later power products. By induction hypothesis, each term $c_{ij} X^j$ in this sum with $X^j \notin S$ can be replaced by a linear combination of lexicographically later elements from S. Doing so we get a representation of X^i which has the required form.

(iii). Since $\bar{B}$ is a factor ring of $K[X]$, the elements of P generate $\bar{B}$ as a vector space over K. Since $\bar{B}$ is Artinian, an appropriate finite subset of P is also sufficient. Now (ii) implies, that the elements from S are a generating system. Assume the elements of S are linearly dependent in $\bar{B}$. Then there is an element $f \in I$ such that each non-zero term of f is a standard monomial multiplied with an element of K. But the lexicographically first non-zero term of f cannot be such a multiple by the very definition of S. This contradiction proves (iii).

(iv). This is a consequence of (iii) and the fact that B is finitely generated and free over A (see [Ma86], Th. 7.10).

10.11. Structure constants and the equations of $\bar{B}$

Since the standard monomials form a vector space basis of $\bar{B}$ over K, there are identities in $\bar{B}$

$$X^r \cdot X^s = \sum_{X^\alpha \in S} c_\alpha^{rs} X^\alpha \quad (X^r, X^s \in S)$$

with uniquely determined constants $c_\alpha^{rs} \in K$, the *structure constants* of $\bar{B}$ over K with respect to S. Similarly, every $X_i \in S'$ can be written in $\bar{B}$ as a linear combination of lexicographically later standard monomials,

$$X_i = \sum_{X^\alpha \in S} c_{i\alpha} X^\alpha$$

with $c_{i\alpha} \in K$ and $c_{i\alpha} = 0$ whenever $X^\alpha \leq X_i$. Define

$$\bar{F}^{rs}(X) := X^r \cdot X^s - \sum_{X^\alpha \in S} c_\alpha^{rs} X^\alpha \in K[X]$$

$$\bar{F}_i(X) := X_i - \sum_{X^\alpha \in S} c_{i\alpha} X^\alpha$$

The polynomials

$$\bar{F}^{rs}(X), \bar{F}_i(X) \text{ with } X^r, X^s \in S, X_i \in S'$$

will be referred to below as the *defining equations of the fibre $\bar{B}$*. This is justified by the fact that the homomorphism of K-algebras

$$K[X] \to \bar{B}$$

mapping each $X_i \in S$ into itself and $X_i \in S'$ into $\sum_{X^\alpha \in S} c_{i\alpha} X^\alpha$ induces an isomorphism

$$\bar{B}_1 := K[X]/(\bar{F}^{rs}(X), \bar{F}_i(X) \mid X^r, X^s \in S, X_i \in S') \quad \to \quad \bar{B}.$$

The latter homomorphism is clearly well-defined in view of the definition of the polynomials $\bar{F}_i$ and the definition of the multiplication in $\bar{B}$. It is surjective since $\bar{B}$ is generated as a K-vector space by the power products $X^r \in S$. To show that this is an isomorphism, it will be sufficient to show that $\bar{B}_1$ is equal to its K-linear subspace $V \subseteq \bar{B}_1$ generated by the standard monomials. In other words, we have to show that every $X^r \in P$ considered as an element of $\bar{B}_1$ is already in V. This is trivially true if X^r is a standard monomial. Moreover, since the polynomials $\bar{F}^{rs}$ are zero in $\bar{B}_1$, the product of every two standard monomials is in V. This implies, that an element of V multiplied with a standard monomial is again an element of V. So if X^r is a product of finitely many standard monomials, then $X^r \in V$. Now let $X^r \in P$ be arbitrary. If X^r isn't a product of standard monomials, then it is divisible by some of the elements from S'. Write

$$X^r = \cdot X^{r'} \cdot X^{r''}$$

where $X^{r'}$ is a product of elements from S' and $X^{r''}$ is a product of standard monomials. Define the *non-standard degree* of X^r to be

$$\text{nsdeg } X^r := \deg X^{r'}$$

the degree of the non-standard part $X^{r'}$ of X^r. The further proof will precede by induction on nsdeg X^r. If nsdeg $X^r = 0$, the claim is already proved. Assume nsdeg $X^r > 0$. Then there is some $X_i \in S'$ dividing X^r. Since the polynomial $\bar{F}_i$ is zero in $\bar{B}_1$, X_i can be written in $\bar{B}_1$ as a linear combination of standard monomials. Multiply by X^r/X_i to get a presentation of X^r as a linear combination of polynomials with non-standard degrees less than nsdeg X^r. Using the induction hypothesis, we see that X^r is a linear combination of elements from V, i.e., $X^r \in V$.

10.12. A standard base for $\bar{I}$

The polynomials

$$\bar{F}^{e_i s}(X), \bar{F}_j(X) \text{ with } X_i, X^s \in S, X_i \cdot X^s \notin S, \text{ and } X_j \in S'$$

form a standard base for the ideal $\bar{I}$. Here e_i denotes the i-th standard unit vector.

Proof. By definition, the elements of the set $\mathbb{N}^N - S$ are just the power products that occur as initial forms of the elements from $\bar{I}$. This set is partially ordered with respect to the divisibility relation "$|$". The minimal elements with respect to this relation are just the generators of the ideal of initial forms of $\bar{I}$. Such a minimal element $X^r \in \mathbb{N}^N - S$ is characterized by the condition that all proper divisors of X^r are no longer in $\mathbb{N}^N - S$. In particular, $X^r = X_i \cdot X^s$ with $X^s \in S$. If X_i is not in S, X^s must be 1, since X_i itself is then already such a minimal element. So we see, the set of elements

$$X_i \cdot X^s, X_j \quad \text{with } X_i, X^s \in S, X_i \cdot X^s \notin S, \text{ and } X_j \in S'$$

is a set of generators for the ideal of initial forms of $\bar{I}$. But this is just the set of initial forms of the generators in question.

10.13. The order of the standard monomials in $\bar{B}$

Proof. Obviously ord $X^r \geq |r|$. Assume this inequality is strict. Then

$$X^r \in (X)^{|r|+1} K[X] + \bar{I}$$

hence

$$X^r \in \bar{I} + \sum_{X^r < X^\alpha \in P} K \cdot X^i.$$

But this means, X^r is an initial form of some element of $\bar{I}$, i.e., $X^r \notin S$, contrary to the assumption.

10.14. Properties of the structure constants of $\bar{B}$

Let $X^r, X^s, X^t, X^u, X^w \in S$. Then:

(i) $c_t^{rs} = 0$, if $X^r X^s > X^t$.

(ii) $c_t^{rs} = c_t^{sr}$ (commutativity).

(iii) $\sum_{\alpha \in S} (c_\alpha^{rs} c_u^{t\alpha} - c_\alpha^{ts} c_u^{r\alpha}) = 0$ (associativity).

(iv) $c_t^{rs} = \begin{cases} 1 & \text{if } r + s = t \\ 0 & \text{otherwise} \end{cases}$, if the product $X^r X^s = X^{r+s}$ is in S.

(v) $c_u^{r+s,t} = \sum_{\alpha \in S} c_u^{r\alpha} c_\alpha^{st}$, if the product $X^r X^s = X^{r+s}$ is in S.

(vi) $c_w^{rs} = c_w^{uv}$, if $X^r X^s = X^u X^v$.

Note that, as a consequence of (v), the structure constants are already completely determined if the values of the constants

$$c_s^{e_i, r} \quad \text{with } i = 1, \ldots, N, \text{ and } X^r, X^s \in S$$

are fixed. Here e_i denotes the i-th unit vector.

Proof. (i) follows from our definition of a standard monomial. The other identities are derived from special cases of the commutative or associative law such as

$$X^r \cdot X^s = X^s \cdot X^r, \quad (X^r \cdot X^s) \cdot X^t = X^r \cdot (X^s \cdot X^t), \quad X^r X^s = X^{r+s}$$

just by writing both sides as a linear combinations of standard monomials and then comparing coefficients.

10.15. Terminology

In that what follows,

$$c := (c_t^{rs})_{r,s,t \in S}$$

denotes the family of structure constants of $\bar{B}$ over K with respect to S,

$$x := (x_t^{rs})_{r,s,t \in S}$$

denotes a family of indeterminates indexed in the same way as the family c, and

$$y := (y_t^{rs})_{r,s,t \in S} \cup (y_{is})_{(i,s) \in S' \times S}$$

denotes a family of indeterminates which is modified as follows:

1. Certain of the indeterminates are replaced by zeros: $y_t^{rs} = 0$ whenever $X^{r+s} \in S$ (even if $r + s = t$ since these indeterminates will represent deviations from the structure constants of the special fibre rather than structure constants).

2. Certain of the indeterminates are identified: $y_t^{rs} = y_t^{sr}$ for arbitrary $r, s, t \in S$.

Given two families $a := (a_t^{rs})_{r,s,t \in S}$ and $b := (b_t^{rs})_{r,s,t \in S}$ of elements from some ring, we will write

$$< a, b > := \sum_{r,s,t \in S} a_t^{rs} \cdot b_t^{rs}$$

Further define

$$Ass_u^{rst}(x) := \sum_{\alpha \in S} x_\alpha^{rs} x_u^{t\alpha} - x_\alpha^{ts} x_u^{r\alpha}$$

$$Ass(x) := (Ass_u^{rst}(x))_{r,s,t,u \in S}$$

$$dAss_u^{rst}(x) := \left(\frac{\partial Ass_u^{rst}(x)}{\partial x_k^{ij}} \right)_{i,j,k \in S}$$

Note that since $Ass_u^{rst}(x)$ is a quadratic form,

$$Ass_u^{rst}(a + b) = Ass_u^{rst}(a) + < dAss_u^{rst}(a), b > + Ass_u^{rst}(b).$$

Finally we will refer to the polynomials

$$F^{rs}(X, y) := X^r \cdot X^s - \sum_{X^\alpha \in S} (c_\alpha^{rs} + y_\alpha^{rs}) \cdot X^\alpha \in K[y, X], \quad X^r, X^s \in S$$

$$F_i(X, y) := X_i - \sum_{X^\alpha \in S} (c_{i\alpha} + y_{i\alpha}) X^\alpha \in K[y, X], \quad X_i \in S'.$$

as to the *defining equations of the deformation space B over A*.

10.16. The germ of the universal family

Define

$$A := (K[y]/(Ass(c+y)))_{(y)}$$

$$B := \bigoplus_{i \in S} A \cdot X^i$$

$$X^r \cdot X^s := \sum_{\alpha \in S} (c_\alpha^{rs} + y_\alpha^{rs}) \cdot X^\alpha$$

Equivalently

$$B := A[X]/(F^{rs}(X,y), F_i(X,y)|X^r, X^s \in S, X_i \in S', X^{r+s} \notin S)$$
$$= A[X]/(F^{e,s}(X,y), F_j(X,y)|X_i, X^s \in S, X_j \in S', X^{r+s} \notin S)$$

where the polynomials F^{rs} and F_i are defined as in 10.15. Then the map

$$f : A \to B, a \mapsto a \cdot 1 (= a \cdot X^0)$$

is up to isomorphism equal to the germ $O_{\mathbb{U},0} \to O_{\mathbb{H},0}$ at $0 \in \mathbb{H}$ of the universal family (see 10.7)

$$f : \mathbb{U} \subseteq \mathbb{P}_{\mathbb{H}}^N := \mathbb{P}_K^N \times_K \mathbb{H} \to \mathbb{H}.$$

Note that there are no relations between the indeterminates of type $y_{i\alpha}$, which corresponds to the fact that every additional dimension of the embedding space results in plenty of additional unobstructed "deformations" which are essentially motions inside this space and don't give any contribution to Schlessinger's T^1.

Proof. We have to show that the definitions of B given above are equivalent. Assume that B is defined according to the first definition as a direct sum with the indicated multiplication. Then B is a commutative A-algebra and there is an A-algebra homomorphism

$$A[X] \to B$$

mapping each variable $X_i \notin S'$ into itself and each $X_i \in S'$ into the linear combination $\sum_{X^\alpha \in S} (c_{i\alpha} + y_{i\alpha}) X^\alpha$. Since B is generated by the standard monomials, this map is surjective. Moreover, from the definition of the map and of the multiplication in B we see that the polynomials F_i and F^{rs} are in the kernel. So there is a surjective A-algebra homomorphism

$$f_1 : B_1 := A[X]/(F^{rs}(X,y), F_i(X,y)|X^r, X^s \in S, X_i \in S') \to B,$$

which we have to prove is injective. Let C be the A-submodule of B_1 generated by the residue classes of the standard monomials $X^\alpha \in S$. Since these residue classes are mapped to A-linearly independent elements of B, they form a free basis of C over A. In particular, the restriction of f_1 to C is an isomorphism, and it will be sufficient to show $B_1 = C$. Since $1 \in C$ and B_1 is generated as an $A[X]$-module by the unit element, it is sufficient to show that C is an $A[X]$-submodule of B_1. In other words, we have to prove, $X_i \cdot C \subseteq C$ for every variable.

Further it is sufficient to prove this inclusion for standard variables $X_i \in S$ only, since each variable from S' can be written in B_1 as a linear combination of standard monomials (the residue classes in B_1 of the polynomials $F_i(X, y)$ are zero). So assume X_i is in S. The inclusion to be proved means that X_i times an arbitrary standard monomial X^s can be written in B_1 as a linear combination of standard monomials. But this is obvious from the fact that the residue class in B_1 of $F^{e_i s}(X, y)$ is zero. We have proved, the definitions of B given above are identical. Note that our argument applies equally if one factors with respect to the ideal generated by the smaller set of polynomials of type $\bar{F}^{e_i s}(X)$.

We want to prove that the homomorphism $A \to B$ above is the germ of the universal family at $0 \in \mathbb{H}$. By construction the special fibre of f is the vector space

$$B/(y)B \cong \bigoplus_{X^\alpha \in S} K \cdot X^\alpha.$$

equipped with the multiplication

$$X^r \cdot X^s := \sum_{X^\alpha \in S} c^{rs}_\alpha \cdot X^\alpha,$$

i.e., $B/(y)B$ is up to isomorphism the local ring $\bar{B}$ of the fibre of ϕ over the point $0 \in \mathbb{H}$. Since B is finite over the local ring A, the fact that $\bar{B}$ is local implies that B is also local, with maximal ideal, say, n. Further, $f : A \to B$ factors

$$f : A \overset{i}{\longrightarrow} A[X] \overset{j}{\longrightarrow} B$$

into a polynomial extension i and a surjection j. We have written f as an embedded deformation germ. To find out which way this germ behaves under base change, let $g : A \to A'$ be any base change homomorphism. Then $f \otimes_A A'$ is a homomorphism

$$f \otimes_A A' : A' = A \otimes_A A' \to B' = B \otimes_A A' = \Big(\sum_{X^\alpha \in S} A \cdot X^\alpha \Big) \otimes_A A' = \sum_{X^\alpha \in S} A' \cdot X^\alpha$$

where the sum signs on the right both denote direct sums. In particular, B' is generated over A' by the standard monomials $X^\alpha \in S$. We are writing here X^α instead of $X^\alpha \otimes 1$ for the natural image of $X^\alpha \in B$ in $B \otimes_A A' = B'$, so that the homomorphism $h : B \to B'$ induced by $g : A \to A'$ maps each standard monomial X^α into itself. The map h applied to the identities $X^r \cdot X^s = \sum_{X^\alpha \in S} (c^{rs}_\alpha + y^{rs}_t) X^\alpha$ yields the identities in B',

$$X^r \cdot X^s = X^r \otimes 1 \cdot X^s \otimes 1$$

$$= \sum_{X^\alpha \in S} g(c^{rs}_\alpha + y^{rs}_t) \cdot (X^\alpha \otimes 1)$$

$$= \sum_{X^\alpha \in S} (c^{rs}_\alpha + g(y^{rs}_t)) X^\alpha.$$

We see, under base change with respect to $g : A \to A'$ the ring B continues to be the same type of direct sum over its base ring, just A is replaced by A' and the multiplication law changes in that each structure constant is replaced by its image under g. Similarly, from the presentation in B of the non-standard variables as linear combinations of standard monomials, we obtain the identities in B',

$$X_i = \sum_{X^\alpha \in S} (c_{i\alpha} + g(y_{i\alpha})) X^\alpha$$

Now let

$$f' : A' \xrightarrow{i'} A'[X] \xrightarrow{j'} B'$$

be an embedded deformation germ such that $f \otimes A'/m' = f_0$. We have to prove that f' can be obtained from f by base change with respect to a uniquely determined base change homomorphism $g : A \to A'$.

The images under j' of the standard monomials $X^\alpha \in S$ form a free generating set of B' over A'. For, the residue classes in $B'/m'B' = \bar{B}$ of these images form a K-vector space basis and B' is free and finite over A' (use 10.10(iii) and [Ma86], Th. 7.10). Since the multiplication law of B' reduces modulo m' to the one of $\bar{B}$, we can write

$$X^r \cdot X^s = \sum_{X^\alpha \in S} (c_\alpha^{rs} + a_t^{rs}) X^\alpha$$

with uniquely determined elements $a_t^{rs} \in m'$. Comparing this with the multiplication law for the embedded deformation germ obtained by base change, we see that the base change homomorphism $g : A \to A'$ we are looking for must satisfy,

(1) $$g(y_t^{rs}) = a_t^{rs} \quad \text{and similarly} \quad g(y_{i\alpha}) = a_{i\alpha}$$

where the elements $a_{i\alpha} \in m'$ are such that in B',

$$X_i = \sum_{X^\alpha \in S} (c_{i\alpha} + a_{i\alpha}) X^\alpha.$$

Now the algebra A is generated over K by (the residue classes of) the elements $y_t^{rs}, y_{i\alpha}$, i.e., these identities show that the homomorphism g is uniquely determined, if it exists.

To prove the existence of g, take the identities (1) as a definition. Too see that g is correctly defined this way, note that there is a homomorphism of K-algebras,

$$K[y] \to A', y_t^{rs} \mapsto a_t^{rs}, y_{i\alpha} \mapsto a_{i\alpha}.$$

Since multiplication in B' is associative, the members of the family $Ass(c + a)$ are all zero, i.e., the homomorphisms factors over $K[y]/(Ass(c + y))$. Since A' is local and the elements $a_t^{rs}, a_{i\alpha}$ are in m', we get a local homomorphism

$$g : A \to A'.$$

By construction, the embedded deformation germ obtained from f by base change with this homomorphism g is up to isomorphism the given germ f'. This proves that the embedded deformation germ f defined above is really the germ of the universal family at the point in question.

10.17. The embedded deformation functor

As above let $\bar{B} := B/mB$ be the special fibre of the germ $f : A \to B$ of the universal family $\mathbb{U} \to \mathbb{H}$ at $0 \in \mathbb{H}$. Consider the functor

$$D_{\bar{B}} : \quad \text{local } K\text{-algebras} \longrightarrow \text{sets}$$

such that

$$D_{\bar{B}}(A')$$

is the set of all isomorphism classes of embedded deformation germs

$$f' : A' \xrightarrow{i'} A'[X] \xrightarrow{j'} B'$$

with $f' \otimes_{A'} A'/m' = f_0$. Note that two elements of $D_{\bar{B}}(A')$ are equal if and only if the kernels of the two surjections j' are equal. In other words, the elements of $D_{\bar{B}}(A')$ can be identified with A'-flat families $\operatorname{Spec} B' \subseteq \mathbb{A}_{A'}^N$, such that the fibres are in affine N-space and such that the special fibre is $\operatorname{Spec} \bar{B} \subseteq \mathbb{A}_K^N$.

Given a local homomorphism $g : A' \to A''$ of local K-algebras, the mapping

$$D_{\bar{B}}(g) : D_{\bar{B}}(A') \to D_{\bar{B}}(A'')$$

is just the tensor product with A'' over A':

$$D_{\bar{B}}(g)(A' \xrightarrow{i'} A'[X] \xrightarrow{j'} B') := A'' \xrightarrow{i' \otimes A''} A''[X] \xrightarrow{j' \otimes A''} B''.$$

The functor $D_{\bar{B}}$ is called *embedded deformation functor* of $\bar{B}$. The elements of the set $D_{\bar{B}}(A)$ are called *embedded deformations* over A of $\bar{B}$ and, in the special case $A = K[\varepsilon]/(\varepsilon^2)$, *embedded first order deformations*. The set $D_{\bar{B}}(K[\varepsilon]/(\varepsilon^2))$ is also called *tangent space* of $D_{\bar{B}}$.

10.18. Relations to the Hilbert scheme point functor

(i) Consider the functor,

$$H(?) := \operatorname{Hom}_{K-\text{schemes}}(\operatorname{Spec}(?), \mathbb{H}) : \quad \text{local } K\text{-algebras} \longrightarrow \text{sets}$$

which is the restriction of the Hilbert scheme point functor to the points with values in a local K-algebra, i.e., $H(?)$ associates with a local K-algebra

A' the set of embedded deformations of all length n subschemes in $\mathbb{P}^N_K$ with parameter space $\operatorname{Spec}(A')$. The embedded deformation functor of $\bar{B}$ can be identified with the subfunctor of $H(?)$,

$$D_{\bar{B}}(A') = \{\alpha : \operatorname{Spec}(A') \to \mathbb{H} \in H(A') | \alpha(m') = 0 \in \mathbb{H}\},$$

of embedded deformations over $\operatorname{Spec}(A')$ having as special fibre the fibre of the universal family at $0 \in \mathbb{H}$. Here m' denotes the special point of the local scheme $\operatorname{Spec}(A')$.

(ii) Since a base change morphism $\alpha : \operatorname{Spec}(A') \to \mathbb{H}$ with $\alpha(m') = 0$ factors uniquely over $\operatorname{Spec}(A) = \operatorname{Spec}(O_{\mathbb{H},0})$ and is defined by the induced morphism $\operatorname{Spec}(A') \to \operatorname{Spec}(A)$, there is an isomorphism

$$\operatorname{Hom}_{\text{local } K\text{- algebras}}(A, ?) \longrightarrow D_{\bar{B}}(?)$$
$$A \to A' \quad \mapsto \quad (A \to A[X] \to B) \otimes_A A'$$

of functors defined on the category of local K-algebras (and local homomorphisms). Here A is as usual the base of the universal germ $f : A \to B$.

(iii) From the isomorphism in (ii). one obtains as a special case an isomorphism

$$\operatorname{Hom}_K(m/m^2, K) \to D_{\bar{B}}(K[\varepsilon]/(\varepsilon^2)).$$

Note that the K-vector space on the left is just the *Zariski tangent space* of $\mathbb{H}$ at 0. The isomorphism is essentially the isomorphism of (ii) with $A' := K[\varepsilon]/(\varepsilon^2)$,

$$\operatorname{Hom}_{\text{local } K\text{-algebras}}(A, K[\varepsilon]/(\varepsilon^2)) \to D_{\bar{B}}(K[\varepsilon]/(\varepsilon^2)).$$

Note that a local K-algebra homomorphism $g : A \to K[\varepsilon]/(\varepsilon^2)$ is uniquely determined by its restriction to the maximal ideal $m \subseteq A$. Since m^2 is mapped to zero under this restriction the map g defines a K-linear map $m/m^2 \to K\varepsilon$. Conversely, every K-linear map $m/m^2 \to K\varepsilon$ comes from an unique local K-algebra homomorphism g. So the vector space of local K-algebra homomorphisms $A \to K[\varepsilon]/(\varepsilon^2)$ can be identified with

$$\operatorname{Hom}_K(m/m^2, K).$$

10.19. The cotangent space of $\mathbb{H}$ at $0 \in \mathbb{H}$

The cotangent space m/m^2 of $\mathbb{H}$ at $0 \in \mathbb{H}$ is canonically isomorphic to

$$\left(\left(\bigoplus_{X^r, X^s, X^t \in S} K \cdot y_t^{rs} \right) \oplus \left(\bigoplus_{(X_i, X^\alpha) \in S' \times S} K \cdot y_{i\alpha} \right) \right) / \left(\sum_{X^r, X^s, X^t, X^u \in S} K \cdot < dAss_u^{rst}(c), x > \right)$$

(as a vector space over K).

Proof. The natural homomorphism $K[y]_{(y)} \to (K[y]/(Ass(c+y)))_{(y)} = A$ induces a K-linear surjection

$$L : \left(\bigoplus_{X^r, X^s, X^t \in S} K \cdot y_t^{rs} \right) \oplus \left(\bigoplus_{(X_i, X^\alpha) \in S' \times S} K \cdot y_{i\alpha} \right) \cong (y)/(y)^2 \to m/m^2.$$

The kernel of this map is just the linear part of the ideal generated by the elements of the family $Ass(c+y)$ modulo $(y)^2$:

$$\text{Ker}(L) = \left(\left(\bigoplus_{X^r, X^s, X^t \in S} K \cdot y_t^{rs} \right) \oplus \left(\bigoplus_{(X_i, X^\alpha) \in S' \times S} K \cdot y_{i\alpha} \right) \right) \cap \left(Ass(c+y), (y)^2 \right)$$

Note that

$$\begin{aligned} Ass_u^{rst}(c+y) &= Ass_u^{rst}(c) + <dAss_u^{rst}(c), y> + Ass_u^{rst}(y) \\ &= <dAss_u^{rst}(c), y> + Ass_u^{rst}(y) \\ &\equiv <dAss_u^{rst}(c), y> \bmod(y)^2 \end{aligned}$$

Here we have used that $Ass_u^{rst}(c) = 0$ (since multiplication in $\bar{B}$ is associative). Therefore,

$$\text{Ker}(L) = \sum_{X^r, X^s, X^t, X^u \in S} K \cdot <dAss_u^{rst}(c), x>$$

as required.

10.20. The tangent space of $\mathbb{H}$ at $0 \in \mathbb{H}$

The isomorphism 10.18(iii),

$$\text{Hom}_K(m/m^2, K) \to D_{\bar{B}}(K_\varepsilon), \quad K_\varepsilon := K[\varepsilon]/(\varepsilon)^2,$$

is given by

$$L \mapsto (K_\varepsilon \to K_\varepsilon[X] \to B_L)$$

where

$$B_L := K_\varepsilon[X]/I_L$$

and the ideal I_L is generated by the polynomials

$$\bar{F}^{e_1, s}(X) + \varepsilon \sum_{X^\alpha \in S} L(y_\alpha^{e_1, r}) \cdot X^\alpha \quad (X^r, X_i \in S)$$

$$\bar{F}_i(X) + \varepsilon \sum_{X^\alpha \in S} L(y_{i\alpha}) \cdot X^\alpha \quad (X_i \in S')$$

Here e_i denotes the i-th unit vector and we have identified m/m^2 with the vector space of 10.19.

Proof. The linear form $L : m/m^2 \to K$ corresponds to the K-algebra homomorphism

$$A \to K_\varepsilon = K \oplus K\varepsilon, \quad u + v \mapsto u + L(v \bmod m^2)\varepsilon, \quad u \in K(\subseteq A), v \in m.$$

Note that since K maps surjectively to the residue class field of A, each $a \in A$ can be uniquely written as a sum $a = u + v$ with $u \in K(\subseteq A)$ and $v \in m$. The image of L in $D_{\bar{B}}(K_\varepsilon)$ under this homomorphism is

$$(A \to A[X] \to B) \otimes_A K_\varepsilon = K_\varepsilon \to K_\varepsilon[X] \to B \otimes_A K_\varepsilon.$$

So it will be sufficient to show that

$$B_L \cong B \otimes_A K_\varepsilon.$$

Obviously,

$$B \otimes_A K_\varepsilon = K_\varepsilon[X]/J_L = \bigoplus_{\alpha \in S} K_\varepsilon \cdot X^\alpha$$

where J_L is the ideal generated by the natural images in $K_\varepsilon[X]$ of the polynomials

$$F^{e_i s}(X, y), \quad F_i(X, y).$$

These natural images are

$$L_*(F^{rs}) = X^r \cdot X^s - \sum_{\alpha \in S} (c_\alpha^{rs} + \varepsilon L(y_\alpha^{rs})) \cdot X^\alpha = \bar{F}^{rs} + \varepsilon \sum_{\alpha \in S} L(y_\alpha^{e_i, r}) \cdot X^\alpha$$

$$L_*(F_i) = X_i - \sum_{X^\alpha \in S} (c_{i\alpha} + \varepsilon L(y_{i\alpha})) \cdot X^\alpha = \bar{F}_i(X) + \varepsilon \sum_{X^\alpha \in S} L(y_{i\alpha}) \cdot X^\alpha$$

So the ideal J_L is generated by the polynomials $L_*(F^{rs}), L_*(F_i)$ as required.

10.21. The minimal flatifying filtration

Let F be the filtration of A generated over the natural filtration by (the residue classes of) the variables $y_t^{rs}(X^r, X^s, X^t \in S)$ such that

$$\operatorname{ord} y_t^{rs} \geq |r| + |s| - |t|.$$

Then F is the minimal flatifying filtration for the universal germ $A \to B$ (with respect to the natural filtrations).

Proof. This is just a consequence of the fact that by 10.13, $\operatorname{ord} X^r = |r|$ in $\bar{B}$, hence in B, for every $X^r \in S$. For, this implies that the maps

$$b : S \to B, X^r \mapsto X^r \quad \text{and} \quad d : S \to \mathbb{N}, X^r \mapsto |r|$$

constitute an F_B-distinguished base of B over A with F_B the natural filtration of B, so that the claim follows from 7.5 and 7.9.

10.22. Normal module and first order deformations

Let $\bar{B} = K[X]/\bar{I}$ and

$$\bar{f} := (\bar{f}_1, \ldots, \bar{f}_k) \in K[X]^k$$

be a generating system of the ideal $\bar{I}$,

$$\bar{I} = \bar{f} \cdot K[X].$$

Then

(i) The following homomorphism is bijective.

$$N_{\bar{I}} := \mathrm{Hom}_{K[X]}(\bar{I}, \bar{B}) \longrightarrow D_{\bar{B}}(K_\varepsilon), g \mapsto (K_\varepsilon \rightarrow K_\varepsilon[X] \rightarrow K_\varepsilon[X]/I_g)$$

Here $I_g := (\bar{f}_i + \varepsilon \cdot \breve{g}(\bar{f}_i)|i = 1, \ldots, k) K_\varepsilon[X]$ and $\breve{g}(\bar{f}_i)$ denotes some representative in $K_\varepsilon[X]$ of $g(\bar{f}_i) \in \bar{B}$.

(ii) If $\bar{f}$ is a standard base of $\bar{I}$ (with respect to the (X)-adic filtration of $K[X]$), then the following conditions are equivalent.

    a) The elements of $D_{\bar{B}}(K_\varepsilon)$ are tangentially flat.

    b) $\mathrm{ord}_{\bar{B}}\, g(\bar{f}_i) \geq \mathrm{ord}\,\bar{f}_i - 1$ for every $g \in N_{\bar{f}}$.

Here the order on the right is taken with respect to the (X)-adic filtration.

Remark

The ideal I_g of assertion (i) above generated by the elements $\bar{f}_i + \varepsilon \cdot \breve{g}(\bar{f}_i)$ is independent upon the special choice of the representatives $\breve{g}(\bar{f}_i)$, i.e., the map of (i) is well defined.

Proof of the remark. Let $\{\breve{g}(\bar{f}_1), \ldots, \breve{g}(\bar{f}_k)\}$ and $\{\breve{g}'(\bar{f}_1), \ldots, \breve{g}'(\bar{f}_k)\}$ two sets of representatives and

$$I := (\bar{f}_i + \varepsilon \cdot \breve{g}(\bar{f}_i)|i = 1, \ldots, k)$$
$$I' := (\bar{f}_i + \varepsilon \cdot \breve{g}'(\bar{f}_i)|i = 1, \ldots, k)$$

the corresponding ideals. We want to show these ideals are equal. For this, it will be sufficient to show

$$\bar{f}_i + \varepsilon \cdot \breve{g}'(\bar{f}_i) \in I$$

for every i. Since $\breve{g}'(\bar{f}_i)$ and $\breve{g}(\bar{f}_i)$ represent the same element in $\bar{B}$, $\breve{g}'(\bar{f}_i) - \breve{g}(\bar{f}_i) \in \bar{I} = (I, \varepsilon)$, hence $\breve{g}'(\bar{f}_i) - \breve{g}(\bar{f}_i) = f' + \varepsilon f''$ with $f' \in I$ and $f'' \in K[X]$. Substituting $\breve{g}'(\bar{f}_i) = \breve{g}(\bar{f}_i) + f' + \varepsilon f''$ and using $\varepsilon^2 = 0$ we see that

$$\bar{f}_i + \varepsilon \cdot \breve{g}'(\bar{f}_i) = \bar{f}_i + \varepsilon \cdot \breve{g}(\bar{f}_i) + \varepsilon \cdot f' \in I.$$

Proof of 10.22. (i). The elements of $D_{\bar{B}}(K_\varepsilon)$ are flat homomorphisms of the form

$$(1) \qquad\qquad K_\varepsilon \to K_\varepsilon[X] \to K_\varepsilon[X]/I_g$$

with some ideal I_g and with special fibre $\bar{B} = K[X]/\bar{f}K[X]$. Therefore I_g is generated by elements of type

$$\bar{f}_i + \varepsilon\bar{g}_i$$

with $\bar{g}_i \in K[X]$. Note that, by construction, $I_g = (\bar{f}_i + \varepsilon\bar{g}_i | i = 1, \ldots, k) + (\varepsilon) \cap I_g$, and by Tjurina's criterion 2.10, the intersection $(\varepsilon) \cap I_g$ can be replaced by the product $(\varepsilon)I_g$, which in turn can be omitted at all by Nakayama's lemma 2.1, i.e., I_g can be generated by elements of the indicated type. Again by Tjurina's criterion 2.10, flatness of (1) now means that each relation of the $\bar{f}_i$'s in $K[X]$ can be lifted to a relation of the elements $\bar{f}_i + \varepsilon\bar{g}_i$ in $K_\varepsilon[X]$, i.e., one has an implication

$$< \bar{r}, \bar{f} > = 0 \Rightarrow \exists \bar{s} : 0 = < \bar{r} + \varepsilon\bar{s}, \bar{f} + \varepsilon\bar{g} > \quad (= < \bar{r}, \bar{f} + \varepsilon\bar{g} > + \varepsilon < \bar{s}, \bar{f} >)$$
$$\Leftrightarrow < \bar{r}, \bar{f} + \varepsilon\bar{g} > \in \varepsilon\bar{I}$$
$$\Leftrightarrow < \bar{r}, \bar{g} > \in \bar{I}$$

where $\bar{g} := (\bar{g}_1, \ldots, \bar{g}_k)$ and $\bar{r} := (\bar{r}_1, \ldots, \bar{r}_k)$ have coordinates in $K[X]$. Note that $\varepsilon\bar{I}$ is an ideal in $K_\varepsilon[X]$. By the above implication there is a well-defined map $g \in N_{\bar{I}}$ with

$$g(\bar{f}_i) := (\bar{g}_i \bmod \bar{I}).$$

This proves, the map of (i) is surjective. To prove injectivity, suppose we have two maps $g, g' \in N_{\bar{I}}$ defining one and the same element of $D_{\bar{B}}(K_\varepsilon)$. Let $\check{g}_i$ and $\check{g}'_i$ denote a representative of $g(\bar{f}_i)$ and $g'(\bar{f}_i)$, respectively. Then the ideals

$$I = (\bar{f}_i + \varepsilon\check{g}_i | i = 1, \ldots, k) \text{ and } I' = (\bar{f}_i + \varepsilon\check{g}'_i | i = 1, \ldots, k)$$

are equal. In particular, $\bar{f}_i + \varepsilon\check{g}_i \in I'$, hence $\varepsilon(\check{g}_i - \check{g}'_i) \in I' \cap (\varepsilon) = \varepsilon I' = \varepsilon\bar{I}$ (use 2.10(ii) and the fact that $K_\varepsilon \to K_\varepsilon[X]/I'$ is flat). There is an element $h_i \in \bar{I}$ with

$$\varepsilon(\check{g}_i - \check{g}'_i - h_i) = 0.$$

Since the sum $K_\varepsilon[X] = K[X] + \varepsilon K[X]$ is direct, we deduce $\check{g}_i - \check{g}'_i = h_i \in \bar{I}$, i.e., $g(\bar{f}_i) = g'(\bar{f}_i)$ for every i, thus $g = g'$.

(ii). The condition of (ii)b) is equivalent to

$$\mathrm{ord}(\bar{f}_i + \varepsilon \cdot \check{g}(\bar{f}_i)) = \mathrm{ord}\, \bar{f}_i \text{ for every } i$$

for appropriate lifts $\check{g}(\bar{f}_i)$. In order words, a standard base of $\bar{I}$ can be lifted to a standard base of I_g such that all orders are preserved. But this is, by 4.6(iv), equivalent to tangential flatness of the flat local homomorphism $K_\varepsilon \to K_\varepsilon[X]/I_g$.

10.23. Formal degrees

To each variable x_t^{rs} we associate an integer,

$$d(x_t^{rs}) := |r| + |s| - |t|$$

which we call the *formal degree* of x_t^{rs}. Similarly,

$$d(y_t^{rs}) := |r| + |s| - |t| \text{ and } d(c_t^{rs}) := |r| + |s| - |t|$$

will be called the formal degrees of y_t^{rs} and c_t^{rs}, respectively. Note that the formal degree of, say, c_t^{rs} is a function of the triple (r, s, t), rather than of the field element c_t^{rs}. Structure constants which are equal as field elements may nevertheless have different formal degrees. Therefore, in the context of formal degree, the elements c_t^{rs} will be temporarily treated as if they were indeterminates. An entry as above with formal degree > 1 will be called *critical*. An entry with negative formal degree will be called *irrelevant*. Further we will speak sometimes of the formal degree of a polynomial in the obvious sense.

10.24. Properties of the formal degree

(i) The equations of the associative law $Ass_u^{rst}(x)$ are homogeneous with respect to formal degree. Their formal degrees are

$$d(Ass_u^{rst}(x)) = |r| + |s| + |t| - |u|$$

(ii) Each structure constant c_t^{rs} is either zero or has non-positive formal degree:

$$d(c_t^{rs}) \leq 0 \quad \text{if} \quad c_t^{rs} \neq 0.$$

(iii) Let $G(\bar{B})$ be the associated graded ring with respect to the natural filtration of $\bar{B}$. Then the structure constants of $G(\bar{B})$ over K are obtained from the structure constants c_t^{rs} of $\bar{B}$ over K replacing the irrelevant entries by zero.

Proof. (i). Follows from the very definition of Ass(x),

$$Ass_u^{rst}(x) := \sum_{X^\alpha \in S} x_\alpha^{rs} x_u^{t\alpha} - x_\alpha^{ts} x_u^{r\alpha}.$$

(ii). Follows from property 10.14(i) of the structure constants.

(iii). Let c^0 denote the family of structure constants of $G(\bar{B})$ over K with respect to the (initial forms of the) standard monomials $X^r \in S$. Recall that these initial forms are by 5.3 free generators of $G(\bar{B})$ over K since by 10.10(iii) and 10.13 the maps

$$S \to B, X^r \mapsto X^r \quad \text{and} \quad S \to \mathbb{N}, X^r \mapsto |r|$$

constitute an $F_{\bar{B}}$-distinguished basis of $\bar{B}$ over K with $F_{\bar{B}}$ equal to the natural filtration of $\bar{B}$. Consider the defining identities in 10.11 for the structure constants c. Note that these identities are homogeneous if the standard monomials are equipped with the usual and the structure constants with the formal degree (by 10.13). In passing to the initial forms with respect to the natural filtration on both sides one can omit all terms on the right with a coefficient of negative formal degree, for, such terms have a (usual) degree greater that the degree of the initial term. Therefore the family c^0 is obtained from c replacing by zero all structure constants of negative formal degree. Since the structure constants of positive formal degree are zero from the very beginning (by (ii)), the non-zero elements in c^0 have formal degree zero.

10.25. First order deformations and critical variables

The following conditions are equivalent.

(i) The first order deformations of $\bar{B}$ are tangentially flat.

(ii) $G(N_{\bar{I}})$ doesn't have any non-zero homogeneous elements of degree < -1. Here the normal module $N_{\bar{I}}$ is equipped with the filtration coming from the natural filtrations on $\bar{I}$ and $\bar{B}$.

(iii) The residue classes in A/m^2 of the critical variables y_t^{rs} are zero.

Proof. (i)⇔(ii). Condition (ii) says,

(1) $\quad \operatorname{ord}_{\bar{B}} g(f) \geq \operatorname{ord}_{(X)}(f) - 1$ for arbitrary $f \in \bar{I}, g \in N_{\bar{I}} = \operatorname{Hom}_{K[X]}(\bar{I}, \bar{B})$.

By 10.22(ii) b) it will be sufficient to show that this is equivalent to

(2) $$\operatorname{ord}_{\bar{B}} g(\bar{f}_i) \geq \operatorname{ord}_{(X)}(\bar{f}_i) - 1$$

for every element $\bar{f}_i$ of a standard base $\bar{f} := (\bar{f}_1, \ldots, \bar{f}_k)$ of $\bar{I}$. The implication (1)⇒(2) is trivial. Assume (2) is satisfied and let f be an element of order d in $\bar{I}$. Then, since $\bar{f}$ is a standard base,

$$f = < \bar{f}, r >$$

for some $r = (r_1, \ldots, r_k) \in K[X]$ satisfying $\operatorname{ord}_{(X)} r_i \geq d - \operatorname{ord}_{(X)} \bar{f}_i$ for every i. But then $g(f) = \sum_i r_i g(\bar{f}_i)$, hence for at least one i,

$$\operatorname{ord} g(f) \geq \operatorname{ord} r_i + \operatorname{ord} g(\bar{f}_i) \geq \operatorname{ord} r_i + \operatorname{ord} \bar{f}_i - 1 \geq d - 1 = \operatorname{ord} f - 1.$$

(iii)⇒(i). Let $L : m/m^2 \to K$ be a linear form. We have the show that the first order deformation defined by L is tangentially flat. By assumption,

(3) $$L(y_t^{rs}) = 0 \quad \text{if} \quad y_t^{rs} \text{ is critical.}$$

The first order deformation defined by L has an ideal with basis

$$\bar{F}^{e_i,s}(X) + \varepsilon \sum_{X^\alpha \in S} L(y_\alpha^{e_i,r}) \cdot X^\alpha \quad (X^r, X_i \in S)$$

$$\bar{F}_i(X) + \varepsilon \sum_{X^\alpha \in S} L(y_{i\alpha}) \cdot X^\alpha \quad (X_i \in S')$$

(see 10.20). Condition (3) implies that the order of these polynomials is preserved when ε is replaced by zero. In other word, it is possible the lift a standard basis of $\bar{I}$ preserving the orders of the equations. But this means the corresponding deformation is tangentially flat (see 10.12 and 4.6(iv)).

(i)$\Rightarrow$(iii). Assume there is a critical variable y_t^{rs} having a non-zero residue class in m/m^2. Then this residue class is part of K-vector space basis of m/m^2 and there is a linear form $L : m/m^2 \to K$ with $L(y_t^{rs}) \neq 0$. It will be sufficient to show that

$$B_L := K_\varepsilon[X]/I_L \text{ with } I_L := (L_*(F^{rs}(X,y), L_*(F_i(X,y))|X^r, X^s \in S, X_i \in S')$$

is not tangentially flat over K_ε.

Assume, on the contrary, it is. Since $L_*(F^{rs}(X,y))$ is a lift to I_L of the standard generator $\bar{F}^{rs}(X) \in \bar{I}$, it is uniquely determined modulo I_L , i.e.,

$$L_*(F^{rs}(X,y) - \bar{F}^{rs}(X)) = \varepsilon \cdot \sum_{X^\alpha \in S} L(x_\alpha^{rs})X^\alpha$$

is modulo I_L independent upon the special choice of the lift of $\bar{F}^{rs}(X)$.

Since B_L is tangentially flat over K_ε, at least one lift of $\bar{F}^{rs}(X)$ has the same order like $\bar{F}^{rs}(X)$. This gives an estimation for the order in the factor ring B_L,

$$\text{ord}_{B_L}(\varepsilon \cdot \sum_{X^\alpha \in S} L(x_\alpha^{rs})X^\alpha) \geq \text{ord}_{(X)}\bar{F}^{rs}(X) \geq |r| + |s|.$$

Since ε has order 1 in K_ε, tangential flatness of B_L over K_ε implies (see 5.10),

$$\text{ord}_{\bar{B}}(\sum_{X^\alpha \in S} L(x_\alpha^{rs})X^\alpha) \geq |r| + |s| - 1,$$

hence in $K[X]$,

(4) $$\sum_{X^\alpha \in S} L(x_\alpha^{rs})X^\alpha \in (X)^{|r|+|s|-1} + \bar{I}.$$

Now $L(y_t^{rs}) \neq 0$ and the variable y_t^{rs} is a critical one, i.e.,

$$|t| < |r| + |s| - 1.$$

Therefore, the degreewise lexicographically first α with $L(y_\alpha^{rs}) \neq 0$ satisfies the analogue inequality $|\alpha| < |r| + |s| - 1$. But then (4) implies, that X^α is the initial form of an element of $\bar{I}$, contradicting the fact that X^α is a standard monomial.

10.26. Tangential flatness and the normal module

Equip $K[X]$ with the (X)-adic filtration (which induces the natural filtration on $\bar{B} = K[X]/\bar{I}$) and consider

$$N_{\text{in}(\bar{I})} := \text{Hom}_{K[X]}(\text{in}(\bar{I}), \text{G}(\bar{B})).$$

Assume that this module doesn't have any non-zero homogeneous elements of degree < -1. Then the universal germ

$$A \to A[X] \to B$$

is tangentially flat.

Remark 1

The assertion of 10.26 follows of course directly from our main result 9.2. The assumption on the normal module, which can be considered as an assumption on Schlessiger's T^1, implies

$$\text{H}_A^1 \cdot \text{H}_{B/mB}^0 \leq \text{H}_B^1 \cdot 1 \quad (\leq \text{H}_A^1 \cdot \text{H}_{B/mB}^0 \text{ by } 6.13)$$

so that equality holds and B is tangentially flat over A. Below we give an alternative proof of 10.26 that uses the terminology of this chapter. Note that tangential flatness of the universal germ $A \to K[X] \to B$ implies that every (K-rational) deformation of $\bar{B}$ is tangentially flat. So this reestablishes (in the K-rational case and for zero dimensional fibres) the main result of [He91].

Proof of the tangential flatness of $A \to B$. Step 1: We prove that the first order deformations of $\text{G}(\bar{B})$ are tangentially flat.

Define

$$\text{G}(\bar{B}) := \prod_{n=0}^{\infty} \text{G}(\bar{B})(n) = \lim_n \text{proj} \, \text{G}(\bar{B})/(\text{G}^+(\bar{B}))^n$$

to be the completion of $\text{G}(\bar{B})$ at the irrelevant ideal. This is a local ring, which can be written as a factor ring of $K[[X]]$ modulo the ideal generated by the initial forms of the elements from $\bar{I}$,

$$\text{G}(\bar{B}) \cong K[[X]]/(\text{in}(\bar{I})).$$

We want to show that the first order deformations of $G(\bar{B})$ are tangentially flat. For this, consider the normal module

$$N_{(\text{in}(\bar{I}))} := \text{Hom}_{K[[X]]}((\text{in}(\bar{I})), G(\bar{B})).$$

The usual degree filtrations on $(\text{in}(\bar{I}))$ and $G(\bar{B})$ define a filtration F on $N_{(\text{in}(\bar{I}))}$ with F^d consisting of those maps that shift the initial degrees of the elements by at least d. Looking at the homogeneous elements of $N_{(\text{in}(\bar{I}))}$ one sees that the associated graded module can be identified with

$$G(N_{(\text{in}(\bar{I}))}) = \text{Hom}_{K[X]}(\text{in}(\bar{I}), G(\bar{B})) = N_{\text{in}(\bar{I})}.$$

Thus, by assumption, the module $G(N_{(\text{in}(\bar{I}))})$ doesn't have any non-zero homogeneous elements of degrees < -1, i.e., by 10.25, the first order deformations of $G(\bar{B})$ are tangentially flat. Since $\bar{B}$ has finite length, $G(\bar{B}) = G(\bar{B})$. This completes the proof of the first step.

Step 2: We prove that for every critical variable y_t^{rs} there is a polynomial $p_t^{rs}(y)$ which is homogeneous of degree $d(y_t^{rs}) = |r| + |s| - |t|$ with respect to formal degree such that

(1) $$y_t^{rs} \equiv p_t^{rs}(y) \bmod Ass(c^0 + y)$$

and such that each term of $p_t^{rs}(y)$ has usual degree $\geq d(y_t^{rs})$.

In the first step we have proved that the first order deformations in

$$D_{G(\bar{B})}(K_\epsilon)$$

are tangentially flat. So the critical variables have (usual) order ≥ 2 in

$$K[y]_{(y)}/(Ass(c^0 + y)).$$

(by 10.25(iii) and 10.24(iii)). In other words, for every critical variable y_t^{rs} there is a polynomial $p_t^{rs}(y)$ of usual degree ≥ 2 such that (1) is satisfied. The polynomials of the family $Ass(c^0 + y)$ are homogeneous in y and c^0 with respect to formal degree (see 10.24(i)), and all terms have usual degrees ≥ 2. In particular, we may assume that the polynomials $p_t^{rs}(y)$ are homogeneous with respect to formal degree. Moreover, if a critical variable occurs in $p_t^{rs}(y)$, we can use the congruences (1) to replace it by a polynomial. This is a homogeneous substitution with respect to formal degree but replaces something linear by a sum of terms of usual degrees ≥ 2. So if $p_t^{rs}(y)$ has terms of usual degrees $< d(p_t^{rs})$, the number of terms of lowest degree decreases. After a finite number of such substitutions each term of p_t^{rs} has usual degree $\geq d(p_t^{rs})$ (which is automatically the case if there are no more critical variables).

Step 3: We prove, the structure constants of B satisfy

(2) $$\text{ord}_A(c_t^{rs} + y_t^{rs}) \geq |r| + |s| - |t|.$$

10.26

In particular, $A \to B$ is tangentially flat (by 5.8(iii)).

The identities of the previous step give a congruence

$$(3) \qquad y_t^{rs} \equiv p_t^{rs}(y) + q_t^{rs}(y, c) \bmod Ass(c + y)$$

for every critical variable y_t^{rs}. Here $q_t^{rs}(y, c)$ is a polynomial which is such that a irrelevant structure constant occurs in each term and hence is zero modulo the irrelevant structure constants. Moreover, with respect to formal degree this polynomial is homogeneous of (formal) degree $|r| + |s| - |t|$ when considered as a polynomial in the variables both y and c. In particular, each term has a formal degree $> |r| + |s| - |t|$, when considered as a monomial in the variables of y only (since the structure constants have formal degrees ≤ 0 and at least one constant with strictly negative formal degree occurs). Below we will speak about a p-term of the right hand side of (3) if it is a term occurring in $p_t^{rs}(y)$. The terms of $q_t^{rs}(y, c)$ will be called q-terms.

We can now continue the substitution process described in Step 2. Whenever a critical variable occurs in $q_t^{rs}(y, c)$ we can use the congruences (3) to replace it by a sum of terms which have (with respect to y only) either a higher usual degree (if a p-term is involved) or a higher formal degree (if so is a q-term). So if there is a term in $q_t^{rs}(y, c)$ of usual degree less than $|r| + |s| - |t|$, we can carry out a substitution such that the sum of usual and formal degree increases. But the formal degree cannot increase ad infinitum without any increase of the usual degree (since there are only finitely many variables in y, i.e., there is an upper bound for the numbers $d(y_t^{rs})$). We conclude that, after a finite number of steps, each term of $p_t^{rs}(y) + q_t^{rs}(y, c)$ has usual degree $\geq |r| + |s| - |t|$. But this is just the claim (2).

Remark 2

From the results above we get implications

$(1) \Rightarrow (2) \Rightarrow (3)$

where

(1) $N_{in(\bar{I})}$ doesn't have any non-zero homogeneous elements of degree < -1.

(2) The K-rational deformations of $\bar{B} = K[X]/\bar{I}$ are tangentially flat.

(3) $G(N_{\bar{I}})$ doesn't have any non-zero homogeneous elements of degree < -1, i.e., the first order deformations of $\bar{B}$ are tangentially flat.

The statement that follows gives a condition equivalent to (2) and essentially requiring that the order d deformations of $\bar{B}$ are tangentially flat for a well-defined number d.

168

10.27. Tangential flatness and higher order deformations

Let $\bar{f} := (\bar{f}_1, \ldots, \bar{f}_k)$ be a standard basis of the ideal $\bar{I}$ defining $\bar{B} = K[X]/\bar{I}$. Assume that

$$\operatorname{ord} \bar{f}_i \le d \text{ for every } i.$$

Then the following conditions are equivalent.

(i) The universal germ $A \to K[X] \to B$ is tangentially flat.

(ii) The local homomorphism $A/m^d \to B/m^d B$ is tangentially flat.

(iii) The elements of $D_{\bar{B}}(\bar{A})$ are tangentially flat for every factor ring $\bar{A}$ of $K[T]/(T)^d$. Here $T = (T_1, \ldots, T_e)$ are indeterminates and $e := \dim_K N_{\bar{I}}$.

In checking condition (iii) one may restrict to rings $\bar{A}$ that are Gorenstein.

Proof. (i)$\Rightarrow$(ii). Follows from 3.8(i).

(ii)$\Rightarrow$(iii). The elements of $D_{\bar{B}}(\bar{A})$ are obtained from the universal germ $A \to K[X] \to B$ by base change with respect to $g : A \to \bar{A}$, where g is a residually rational homomorphism of local K-algebras. Since $\bar{A}$ is assumed to be a factor ring of $K[T]/(T)^d$, the base change homomorphism $g : A \to \bar{A}$ factors over the canonical homomorphism $A \to A/m^d$. In other words, the elements of $D_{\bar{B}}(\bar{A})$ are obtained from $f \otimes A/m^d : A/m^d \to B/m^d B$ by residually rational base change. Hence tangential flatness of $f \otimes A/m^d$ implies that the elements of $D_{\bar{B}}(\bar{A})$ are tangentially flat by 3.17.

(iii)$\Rightarrow$(ii). This is trivial since A/m^d is a factor ring of $K[T]/(T)^d$ if the number of indeterminates T_i is at least

$$\dim_K m/m^2 = \dim_K \operatorname{Hom}_K(m/m^2, K) = \dim_K N_{\bar{I}}$$

(see 10.18(iii) and 10.22(i))

(ii)$\Rightarrow$(i). Assume the universal germ $f : A \to B$ is not tangentially flat. Write $\bar{B} = K[X]/\bar{I}$ and consider the standard basis of the ideal $\bar{I}$ formed by the polynomials $\bar{F}^{rs}(X)$ with $X^r, X^s \in S$ and $\bar{F}_j(X)$ with $X_j \in S'$. By assumption, $\bar{I}$ has a standard base consisting of polynomials with initial degrees $\le d$. Therefore, the polynomials

$$\bar{F}^{rs}(X), \bar{F}_j(X) \quad \text{with } X^r, X^s \in S, X_j \in S' \text{ and } |r| + |s| \le d$$

also form a standard basis of $\bar{I}$. Now consider the lifts $F^{rs}(X,y)$, $\bar{F}_j(X,y)$ of these generators to elements of the defining ideal of B,

$$F^{rs}(X,y) = X^r \cdot X^s - \sum_{X^\alpha \in S} (c_\alpha^{rs} + y_\alpha^{rs}) \cdot X^\alpha \in A[[X]],$$

$$F_i(X,y) = X_i - \sum_{X^\alpha \in S} (c_{i\alpha} + y_{i\alpha}) X^\alpha \in A[[X]],$$

$(X^r, X^s \in S, |r| + |s| \le d$ and $X_i \in S')$. For at least one such lift we have

$$\operatorname{ord}_{(m,X)} F^{rs}(X,y) < \operatorname{ord}_{(X)} \bar{F}^{rs}(X) = |r| + |s| (\le d),$$

for otherwise, $A \to B$ were tangentially flat by 4.6(iv). Note that the order of $\bar{F}_j(X)$ (which is equal to 1) cannot decrease when the generator is lifted. So, for at least one pair (r,s),

$$\text{ord}_{(m,X)}\left(\sum_{X^\alpha \in S} y_\alpha^{rs} \cdot X^\alpha \right) = \text{ord}_{(m,X)}(F^{rs}(X,y) - \bar{F}^{rs}(X)) < |r| + |s|(\leq d),$$

hence

$$(1) \qquad\qquad \text{ord}_A(y_t^{rs}) + |t| < |r| + |s| \leq d$$

for at least one t with $X^t \in S$. Here we denote the residue class in A of y_t^{rs} also by y_t^{rs}. It will be sufficient to show, (1) implies that $f \otimes A/I$ cannot be tangentially flat for ideals I with

$$m^{|r|+|s|-|t|} \subseteq I \quad \text{and} \quad y_t^{rs} \notin I,$$

for, this contradicts the assumption that $f \otimes A/m^d$ hence $f \otimes A/I$ is tangentially flat. Note that by (1), for example $I := m^{|r|+|s|-|t|}$ is an ideal satisfying these conditions, i.e., the set of ideals I is not empty . Let

$$a_\alpha^{rs} := (y_\alpha^{rs} \bmod I)$$

denote the residue class in A/I of the element y_α^{rs} of A. Then one has in B/IB the identities

$$\sum_{X^\alpha \in S} (c_\alpha^{rs} + a_\alpha^{rs}) \cdot X^\alpha = X^r \cdot X^s \in n^{|r|+|s|} \cdot B/IB.$$

Since $f \otimes A/I$ is tangentially flat, this implies by 5.8, $c_\alpha^{rs} + a_\alpha^{rs} \in m^{|r|+|s|-|\alpha|} \cdot A/I$ for every α. Now (1) implies $0 \leq \text{ord}_A y_t^{rs} < |r| + |s| - |t|$, so the formal degree of c_t^{rs} is positive, i.e., $c_t^{rs} = 0$. Therefore

$$a_t^{rs} \in m^{|r|+|s|-|t|} \cdot A/I = 0$$

contradicting the assumption $y_t^{rs} \notin I$.

To prove the final remark of 10.27 concerning the Gorenstein property of $\bar{A}$, it will be sufficient to find a factor ring $\bar{A}$ of A/m^d which is Gorenstein and such that $D_{\bar{B}}(\bar{A})$ contains a tangentially non-flat element. As we just have seen, the assumption that $f : A \to B$ is not tangentially flat implies that the same is true for $f \otimes A/I$ with arbitrary ideals I satisfying

$$m^{|r|+|s|-|t|} \subseteq I \quad \text{and} \quad y_t^{rs} \notin I$$

(for certain tuples r, s, t, with $|r| + |s| \leq d$). It is easy to see that one can find an ideal I with these properties such that y_t^{rs} represents a non-zero socle element in A/I. Increasing I if necessary, we may assume that y_t^{rs} generates the socle in A/I, i.e., A/I has a one-dimensional socle and hence is Gorenstein.

References

[Be70]   Bennett, B. M.: On the characteristic functions of a local ring, Ann. Math. 91 (1970) 25-87

[Bo61]   Bourbaki, N.: Algèbre commutative, Hermann, Paris 1961-1965

[Bru85] Brundu, M.: On tangential flatness, Commun. Algebra 13 (1985) 1491-1508

[D-C70] Dieudonné, J.A., Carrel, T.: Invariant theory, old and new, Advances in Math. 4 (1970)

[Del71] Deligne, P.: Théorie de Hodge II, Publ. Math. IHES 40 (1971), 5-58

[Do72]   Долгачев, И. В.:Абстрактная алгебраическая геометрия, Итоги науки, Алгебра Геометрия Топология, т.10 (1972) стр. 47-113

[G-D64] Grothendieck, A., Dieudonné, J.A.: Eléments de géométrie algébrique VI_1, Publ. Math. IHES No. 20, Paris 1964

[G-D71] Grothendieck, A., Dieudonné, J.A.: Eléments de géométrie algébrique, Springer-Verlag, Berlin 1971

[Gr60]   Grothendieck, A.: Techniques de construction et theorèmes d' existence en géométrie algebrique IV: Les schemes de Hilbert, Seminaire Bourbaki 221 (1960/61)

[H-O82] Herrmann, M., Orbanz, U.: Two notes on flatness, Manuscripta Math. 40 (1982) 109 - 133

[Ha66]   Hartshorne, R.: Connectedness of the Hilbert scheme, Publ. Math. IHES 29 (1966) 5-48

[Ha77]   Hartshorne, Algebraic geometry, Springer-Verlag, New York 1977

[He80]   Herzog, B.: Die Wirkung lokaler Homomorphismen auf die Hilbert function, Math. Nachr. 97 (1980) 103-115

[He80'] Herzog, B.: On the Macaulayfication of local rings, J. Algebra 67 (1980) 305-317

[He82]   Herzog, B.: On a relation between the Hilbert functions belonging to a local homomorphism, J. London Math. Soc. 25 (1982), 458-466

[He83]   Herzog, B.: A criterion for tangential flatness, Manuscripta Math. 43 (1983), 219-228

[He90]   Herzog, B.: Hironaka-Lech inequalities for flat couples of locals rings, Manuscripta Math. 68 (1990), 351-371

[He91]   Herzog, B.: Local singularities such that all deformations are tangentially flat, Trans. Amer. Math. 324 (1991), 555-601

[He92]   Herzog, B.: Tangential flatness for filtered modules over local rings, Reports, Department of Mathematics, University of Stockholm 4 (1991) vi+113 pp

[Hi70]   Hironaka, H.: Certain numerical characters of singularities, J. Math. Kyoto Univ. 10 (1970), 151-187

[Ja90]   Jahnel, J.: Tangentiale Flachheit und Lech-Hironaka-Ungleichungen, Promotionsschrift Jena 1990

[Ja93]   Jahnel, J.: Lech's conjecture on deformations of singularities and second Harrison cohomology, to appear in J. London Math. Soc.

[Ko86]   Kodaira, K.: Complex manifolds and deformations of complex struc-
         tures, Springer-Verlag, Tokyo 1986

[L-L81] Larfeldt, T., Lech, C.: Analytic ramification and flat couples of local
         rings, Acta math. 146 (1981) 201-208

[Le?]   Lech, C: Outline of a proof for $H^1(m_0) \leq H^1(m)$ for flat couples of
         local rings (Q_0, Q) with maximal ideals (m_0, m) such that Q/m_0Q is a
         complete intersection, unpublished.

[Le59]   Lech, C.: Note on multiplicities of ideals, Ark. Mat. 4 (1959), 63-86

[Le64]   Lech, C.: Inequalities related to certain couples of local rings, Acta
         Math. 112 (1964),69-89

[Ma86]   Matsumura, H.: Commutative ring theory, Cambridge University Press
         1986

[Mu66]   Mumford, D.: Lectures on Curves on an algebraic surface, Ann. of
         Math. Studies 59, Princeton University Press 1966

[Na55]   Nagata, M.: The theory of multiplicity in general local rings, in: Proc.
         Intern. Symp. Tokyo-Nikko 1955, Science Councel of Japan, 1956,
         191-226

[Pi74]   Pinkham, H. C.: Deformations of singularities with G_m action. Aster-
         isque 20 (1974)

[Po86]   Popov, V. L.: Modern developments in invariant theory, Proceedings of
         the International Congress of Mathematicians (Berkeley, Calif., 1986),
         394-406

[Ri93]   Richter, T.: Lech inequalities for deformations of singularities defined
         by power products of degree 2, Preprint, Jena University

[Sc68]   Schlessinger, M.: Functors of Artin rings, Trans. Amer. Math. Soc.
         130 (1968) 208-222

[Si74]   Singh, B.: Effect of permissible blowing up on the local Hilbert func-
         tions, Invent Math. 26 (1974) 201-212

[Tj69]   Тюрина, Г. Н.: Локально полууниверсальные плоские дефор-
         мации изолированных особенностей комплексных пространств,
         Изв. Акад. Наук СССР, Сер. Матем. 33(1969), 1026-1058

[Va67]   Vasconcelos, W. V.: Ideals generated by R-sequences, J. Algebra 6
         (1967) 309-316

Bernd Herzog

Matematiska Institutionen

Stockholm's Universitet

S-10691 Stockholm

Sweden

herzog@matematik.su.se

Index

Formula Index

Vol. 1551: L. Arkeryd, P. L. Lions, P.A. Markowich, S.R. S. Varadhan. Nonequilibrium Problems in Many-Particle Systems. Montecatini, 1992. Editors: C. Cercignani, M. Pulvirenti. VII, 158 pages 1993.

Vol. 1552: J. Hilgert, K.-H. Neeb, Lie Semigroups and their Applications. XII, 315 pages. 1993.

Vol. 1553: J.-L- Colliot-Thélène, J. Kato, P. Vojta. Arithmetic Algebraic Geometry. Trento, 1991. Editor: E. Ballico. VII, 223 pages. 1993.

Vol. 1554: A. K. Lenstra, H. W. Lenstra, Jr. (Eds.), The Development of the Number Field Sieve. VIII, 131 pages. 1993.

Vol. 1555: O. Liess, Conical Refraction and Higher Microlocalization. X, 389 pages. 1993.

Vol. 1556: S. B. Kuksin, Nearly Integrable Infinite-Dimensional Hamiltonian Systems. XXVII, 101 pages. 1993.

Vol. 1557: J. Azéma, P. A. Meyer, M. Yor (Eds.), Séminaire de Probabilités XXVII. VI, 327 pages. 1993.

Vol. 1558: T. J. Bridges, J. E. Furter, Singularity Theory and Equivariant Symplectic Maps. VI, 226 pages. 1993.

Vol. 1559: V. G. Sprindžuk, Classical Diophantine Equations. XII, 228 pages. 1993.

Vol. 1560: T. Bartsch, Topological Methods for Variational Problems with Symmetries. X, 152 pages. 1993.

Vol. 1561: I. S. Molchanov. Limit Theorems for Unions of Random Closed Sets. X, 157 pages. 1993.

Vol. 1562: G. Harder, Eisensteinkohomologie und die Konstruktion gemischter Motive. XX, 184 pages. 1993.

Vol. 1563: E. Fabes, M. Fukushima, L. Gross, C. Kenig, M. Röckner, D. W. Stroock, Dirichlet Forms. Varenna, 1992. Editors: G. Dell'Antonio, U. Mosco. VII, 245 pages. 1993.

Vol. 1564: J. Jorgenson, S. Lang, Basic Analysis of Regularized Series and Products. IX, 122 pages. 1993.

Vol. 1565: L. Boutet de Monvel, C. De Concini, C. Procesi, P. Schapira, M. Vergne. D-modules, Representation Theory, and Quantum Groups. Venezia, 1992. Editors: G. Zampieri, A. D'Agnolo. VII, 217 pages. 1993.

Vol. 1566: B. Edixhoven, J.-H. Evertse (Eds.), Diophantine Approximation and Abelian Varieties. XIII, 127 pages. 1993.

Vol. 1567: R. L. Dobrushin, S. Kusuoka, Statistical Mechanics and Fractals. VII, 98 pages. 1993.

Vol. 1568: F. Weisz, Martingale Hardy Spaces and their Application in Fourier Analysis. VIII, 217 pages. 1994.

Vol. 1569: V. Totik, Weighted Approximation with Varying Weight. VI, 117 pages. 1994.

Vol. 1570: R. deLaubenfels, Existence Families, Functional Calculi and Evolution Equations. XV, 234 pages. 1994.

Vol. 1571: S. Yu. Pilyugin, The Space of Dynamical Systems with the C^0-Topology. X, 188 pages. 1994.

Vol. 1572: L. Göttsche, Hilbert Schemes of Zero-Dimensional Subschemes of Smooth Varieties. IX, 196 pages. 1994.

Vol. 1573: V. P. Havin, N. K. Nikolski (Eds.), Linear and Complex Analysis - Problem Book 3 - Part I. XXII, 489 pages. 1994.

Vol. 1574: V. P. Havin, N. K. Nikolski (Eds.), Linear and Complex Analysis - Problem Book 3 - Part II. XXII, 507 pages. 1994.

Vol. 1575: M. Mitrea, Clifford Wavelets, Singular Integrals, and Hardy Spaces. XI, 116 pages. 1994.

Vol. 1576: K. Kitahara, Spaces of Approximating Functions with Haar-Like Conditions. X, 110 pages. 1994.

Vol. 1577: N. Obata, White Noise Calculus and Fock Space. X, 183 pages. 1994.

Vol. 1578: J. Bernstein, V. Lunts, Equivariant Sheaves and Functors. V, 139 pages. 1994.

Vol. 1579: N. Kazamaki, Continuous Exponential Martingales and BMO. VII. 91 pages. 1994.

Vol. 1580: M. Milman, Extrapolation and Optimal Decompositions with Applications to Analysis. XI, 161 pages. 1994.

Vol. 1581: D. Bakry, R. D. Gill, S. A. Molchanov, Lectures on Probability Theory. Editor: P. Bernard. VIII, 420 pages. 1994.

Vol. 1582: W. Balser, From Divergent Power Series to Analytic Functions. X, 108 pages. 1994.

Vol. 1583: J. Azéma, P. A. Meyer, M. Yor (Eds.), Séminaire de Probabilités XXVIII. VI, 334 pages. 1994.

Vol. 1584: M. Brokate, N. Kenmochi, I. Müller, J. F. Rodriguez, C. Verdi, Phase Transitions and Hysteresis. Montecatini Terme, 1993. Editor: A. Visintin. VII. 291 pages. 1994.

Vol. 1585: G. Frey (Ed.), On Artin's Conjecture for Odd 2-dimensional Representations. VIII, 148 pages. 1994.

Vol. 1586: R. Nillsen, Difference Spaces and Invariant Linear Forms. XII, 186 pages. 1994.

Vol. 1587: N. Xi, Representations of Affine Hecke Algebras. VIII, 137 pages. 1994.

Vol. 1588: C. Scheiderer, Real and Étale Cohomology. XXIV, 273 pages. 1994.

Vol. 1589: J. Bellissard, M. Degli Esposti, G. Forni, S. Graffi, S. Isola, J. N. Mather, Transition to Chaos in Classical and Quantum Mechanics. Montecatini, 1991. Editor: S. Graffi. VII, 192 pages. 1994.

Vol. 1590: P. M. Soardi, Potential Theory on Infinite Networks. VIII, 187 pages. 1994.

Vol. 1591: M. Abate, G. Patrizio, Finsler Metrics - A Global Approach. IX, 180 pages. 1994.

Vol. 1592: K. W. Breitung, Asymptotic Approximations for Probability Integrals. IX, 146 pages. 1994.

Vol. 1593: J. Jorgenson & S. Lang, D. Goldfeld, Explicit Formulas for Regularized Products and Series. VIII, 154 pages. 1994.

Vol. 1594: M. Green, J. Murre, C. Voisin, Algebraic Cycles and Hodge Theory. Torino, 1993. Editors: A. Albano, F. Bardelli. VII, 275 pages. 1994.

Vol. 1595: R.D.M. Accola, Topics in the Theory of Riemann Surfaces. IX, 105 pages. 1994.

Vol. 1596: L. Heindorf, L. B. Shapiro, Nearly Projective Boolean Algebras. X, 202 pages. 1994.

Vol. 1597: B. Herzog, Kodaira-Spencer Maps in Local Algebra. XVII, 176 pages. 1994.